PRINCIPLES OF DYNAMICS

PRINCIPLES OF DYNAMICS

BY

RODNEY HILL

A Pergamon Press Book

THE MACMILLAN COMPANY
NEW YORK
1964

THE MACMILLAN COMPANY
60 Fifth Avenue
New York 11, N.Y.

This book is distributed by
THE MACMILLAN COMPANY
pursuant to a special arrangement with
PERGAMON PRESS LIMITED
Oxford, England

Library of Congress Catalog Card Number 64-12274

Set in Monotype Times 10 on 12 pt.
and printed in Great Britain by
the Pitman Press, Bath

CONTENTS

PREFACE

CLASSICAL dynamics remains a key subject in the training of any structural engineer, applied mathematician, or theoretical physicist. For the engineer it is the basis of his entire science; for the mathematician a rare discipline and the gateway to modern theories of continua; for the physicist a prelude to relativistic, statistical, and quantum mechanics. Everyone agrees, on the other hand, that the subject must now be taught in a way radically different from what until lately was traditional. However, in regard to manner and content, opinions diverge according to the eventual specialism, and the several extreme viewpoints have been duly expressed in recent books.

My concern here is to present classical dynamics primarily as an exemplar of scientific theory and method; I am convinced that this is educationally indispensable for students of whichever category. Among other things this demands close attention to epistemology, more particularly the precise meaning and status of each concept or postulate. The familiar long-perpetuated obscurities are today intolerable: echoing a thought of Hertz, we should no longer feel the need to hurry apologetically over fundamentals in the hope that the applications will speak for themselves. Accordingly, I have dealt with the philosophy and logic of Newtonian and Eulerian mechanics more extensively than in other undergraduate texts, and I hope with greater rigour and clarity.

As to content, gravitational field theory is given prominence for reasons both of methodology and the resurgence of interest, popular and technical, in celestial mechanics. Subsequent topics in the general motion of a mass-point and of a rigid body illustrate basic principles and point to current research in geodynamics, space navigation, and stability of control systems; the topics are mostly standard but their treatment has some novelty. To bring perspective to rigid dynamics an introductory account of stress analysis is included, even though a little unusual in a work of this kind. Exercises follow each chapter and either develop the text or encourage students to explore along directed lines; time-honoured problems of no intrinsic merit have been firmly excluded.

An intended second volume will deal with Lagrangian and Hamiltonian mechanics for arbitrary systems. In my experience such generalizations and abstractions are best left until students have some mathematical maturity and sufficient mastery of direct techniques.

This first volume was written during a sabbatical year made possible by the generosity of the University of Nottingham in conferring on me a Professorial Research Fellowship after my relinquishment of the Chair of Applied Mathematics.

R. H.

Cambridge

CHAPTER I

GRAVITATIONAL THEORY OF PLANETARY SYSTEMS

1.1 Introductory Remarks

For philosophical and methodological reasons the subject of dynamics is best entered by way of Newton's gravitational theory.

The primary function of this theory is to synthesize observational data on the motions of the solar system (sun, major and minor planets, satellites and comets) and of stars in the galaxy, and then to predict future movements. During three hundred years numerous implications of the theory have been verified over a rapidly widening range of phenomena and in progressively finer detail. This is a continuing process depending on improvements in mathematical techniques, in instrumental accuracy, and in interpretation of the observations (which often involve, in an intricate manner, various physical hypotheses).

In the solar system there remains just one discordant observation with apparent significance, but even this is quantitatively unimportant except as a long-term effect. It concerns the slow rotation of certain planetary orbits. The perihelion of Mercury, for example, advances at the rate of about 9·6 minutes of arc in a century. Perturbations of the orbit by the other known planets leave 0·7 minutes of arc unaccounted for, and so far only the more general theory of gravitation due to Einstein has yielded a tenable explanation for this and similar residuals.

When reduced to essentials, Newton's postulates have an especially simple structure, as scientific theories go, logically as well as analytically. Although not evident from the time-honoured presentation, their context is essentially kinematical, and ideas of 'force' and 'inertial mass' are not in fact needed. The remaining elements are easily separated and made transparent. In this way one gains a firm conceptual base prior to studying the more complex system of Newtonian laws of motion for non-gravitational phenomena.

1.2 Frame of Reference and Unit of Length

Implicit in the gravitational postulates is the hypothesis (or perhaps convention, if Poincaré's standpoint could be sustained†) that all experimentally derived lengths and angles are to be manipulated according to the rules of euclidean geometry. In this assumed euclidean space a frame of reference must be established. Statements about orbital shapes and velocities are otherwise ambiguous.

A frame of reference is initially specified by a certain rigidly-connected triad of directions, not coplanar; it is equally specified by any other such triad whose position and orientation with respect to the original set do not vary in time. For coordinate axes of reference in the frame it is usual to select arbitrarily an orthogonal, intersecting, triad. Different sets of coordinate axes differ only by a fixed translation and rotation in the given frame, and the equations of coordinate transformation between any two sets have constant coefficients. The verbal distinction between a 'frame' and a particular choice of embedded 'axes' is nevertheless useful, and will be adhered to henceforward.

In celestial mechanics the background frame is determined by stars (or perhaps extra-galactic systems) so remote that no variations have been detected in their apparent angular separations. A frame of reference is said to be **sidereal** when rigid triads of directions fixed in it do not appear to rotate with respect to the specified distant objects. Sidereal frames can differ in regard to their translational motions, but such differences do not affect the appearance of the mutual relative orbit of any pair of bodies.

In terrestrial mechanics the metric measure of length is adopted as the standard of reference by international agreement. Since 1960 the **metre** is defined to be a certain number of wavelengths of a particular radiation from the atom krypton. This new standard can be universally maintained with a consistency far greater than was formerly possible through copies of the original prototype bar of platinum–iridium. The British system is based on a recent definition of the yard as 0·9144 metre and an inch as 2·54 cm, both exactly.

In astronomical work the metre is unsuitably small for general use and so a defined mean radius of the earth's orbit is taken as the

† For an analysis of Poincaré's views, and an examination of inter-relations between geometry and mechanics of the real world, see H. Reichenbach: *The Philosophy of Space and Time* (Dover Publications 1957).

astronomical unit, equal to 1·4953... × 10^8 km (1 kilometre = 0·6215 mile). This precise connexion with the terrestrial unit is irrelevant for celestial mechanics as such, but is needed where it merges with terrestrial mechanics, above all in space travel.

1.3 Time Scale and Units

In science, as well as in civil life, the fundamental unit of time is the **second.** This is indirectly defined via the **tropical year,** which is the interval between one vernal equinox and the next. A **vernal equinox** is the unique instant on March 20 or 21 when the sun's centre travels from south to north across the plane through the earth's equator. Put otherwise, it is when the sun is situated momentarily on the intersection of the equatorial plane with the plane of the ecliptic (which contains the earth's orbit in a sidereal frame); this direction is commonly denoted by ♈ and known as the First Point of Aries.† Since successive intervals between equinoxes vary very slightly a datum year has been adopted (beginning 31 December 1899, at Greenwich mean noon). The second is then defined to be the fraction 1/31,556,925·9747 of the datum tropical year, and the civil day as 60 × 60 × 24 = 86,400 seconds = 1/365·2422... of a tropical year.

For social convenience it is desirable that the civil year should contain a whole number of days, and in addition that the calendar should keep step with the actual seasons. The civil year has therefore been defined to contain either 365 or 366 days, in the well-known sequence (to every fourth an extra day, except for century years not divisible by 400). By this means recurring equinoxes and solstices are maintained perpetually close to the same date and time.

In passing, it may be remarked that the **true solar day**, corresponding to successive solar transits of an observer's meridian, is a varying interval, because of the tilt of the earth's axis to the plane of the ecliptic and the non-uniform angular speed of the sun in its apparent orbit. The **mean solar day**, averaged over a year, is just

† For detailed information about astronomical terms and related matters see R. M. L. Baker, Jr. and M. W. Makemson: *An Introduction to Astrodynamics* (Academic Press 1960) or V. M. Blanco and S. W. McCuskey: *Basic Physics of the Solar System* (Addison-Wesley 1961) or S. W. McCuskey: *Introduction to Celestial Mechanics* (Addison-Wesley 1963).

the civil day already described. The variation about the mean, given by an 'equation of time', has a maximum amplitude of some twenty-eight seconds.

In observatories a different unit of time is used. This is the **sidereal day**, defined to be the interval between successive transits of an observer's meridian by ♈, and is practically equal to the interval between meridian transits by any distant star. Since the earth's orbital and rotational motions happen to have the same sense, the number of transits by ♈ exceeds by one the number of solar transits over a period of a year, very nearly (otherwise expressed, to an observer on the earth the sun recedes eastward annually through all the zodiacal constellations). Hence a sidereal day is close to the fraction 365/366 of a civil or mean solar day. The standard sidereal day is defined to contain $86{,}400/1{\cdot}0027379 = 86{,}164{\cdot}09\ldots$ seconds.

A **sidereal year** is the period of the earth's orbital motion in a sidereal frame. Because of a slow precession of the equinoctial positions around the sun's apparent orbit once every 25,735 years (produced by a conical motion of the earth's axis and equatorial plane: §3.8), the sidereal year exceeds the tropical year by a fraction 1/25,735 approximately, or close to 20 minutes, and equals 365·2564... mean solar days. For this same reason a sidereal day is about $86{,}164/(25{,}735 \times 366) \sim 1/110$ second less than the interval between stellar meridian-transits.

1.4 Keplerian Laws of Planetary Motion

Although the apparent movements of the sun and five of the planets had been studied from antiquity, comprehensive and reliable data was first provided by the (naked-eye) observations of Tycho Brahé towards the end of the sixteenth century. From this data Kepler computed the orbits of the earth and Mars round the sun and, from a study of their properties, stated in 1609 two inductions for planetary motion generally.†

> (i) Relative to any sidereal frame the radius from the sun to the earth, or to any other planet, sweeps out sectorial plane area at a uniform rate.

† A penetrating biography of Kepler is to be found in A. Koestler: *The Sleepwalkers* (Hutchinson 1959).

(ii) Relative to sidereal axes with the sun as origin, each planetary orbit is an ellipse with one focus in the sun.

The phraseology has been modernized and the customary numbering reversed to accord with the historical order of discovery. Kepler was later able to confirm these laws also for Venus, Jupiter and Saturn from telescopic observations of his own. Finally, in 1619, he concluded that

(iii) The squares of the orbital periods of all planets are in the same proportion to the cubes of the major axes of the respective ellipses.

It should be noticed that these statements do not involve the *metric* dimensions of the solar system, which were in any event known only roughly at that time.

Kepler's law (i) asserts that planetary orbits are plane curves against the background of the distant stars. Moreover, as Newton demonstrated, it implies that the acceleration relative to the sun is always heliocentric when observed from a sidereal frame (*Principia*†: Book I, Prop. II). For, if (r, θ) are polar coordinates of a planet in such a frame with the sun as origin, the acceleration component transverse to the radius is $d(r^2\dot{\theta})/rdt$ by the elementary standard formula. And this vanishes since $\frac{1}{2}r^2\dot{\theta}$ is the rate of sweep of sectorial area, which is stated to be uniform, η say.

From (ii) it can be concluded that this heliocentric acceleration is proportional to $1/r^2$ (*Principia:* Book I, Prop. XI; the original proof can probably be dated 1676). This follows, for example, by taking the pedal form of the equation to an ellipse:

$$\frac{l}{p^2} = \frac{2}{r} - \frac{1}{a} \tag{1}$$

where p is the perpendicular on a tangent from the focal origin, $2a$ is the major axis, and $2l$ is the latus rectum. This form is suited to the present purpose since it leads immediately to the relation between orbital speed v and distance r by substitution of

$$\frac{1}{p} = \frac{v}{2\eta} \tag{2}$$

† *Mathematical Principles of Natural Philosophy* (1687), together with *The System of the World.* Cajori's revision of Motte's (1729) translation of the Latin original is available in a convenient two-volume edition (University of California Press 1962).

in (1). For any central orbit the inward radial acceleration f then follows as a function of distance from

$$d(\tfrac{1}{2}v^2) + f\,dr = 0. \tag{3}$$

The usual derivations of (3) obscure the simple fact that it is just the self-evident identity $\mathbf{v}d\mathbf{v}/dt \equiv \dot{\mathbf{v}}d\mathbf{r}/dt$ in differential form. By combining the previous equations we have finally

$$f = \frac{\mu}{r^2} \quad \text{where} \quad \mu = \frac{4\eta^2}{l}. \tag{4}$$

For future reference we note at this point the explicit speed–distance relation

$$\frac{v^2}{\mu} = \frac{2}{r} - \frac{1}{a} \tag{5}$$

from (1), (2) and (4).

Law (iii) concerns the orbital period, τ say, which is just the area of the ellipse divided by the rate of sweep; that is, $\pi ab/\eta$ where $b^2 = al$ and $2b$ is the minor axis. Eliminating l from (4) in favour of τ:

$$f = \frac{\mu}{r^2} \quad \text{where} \quad \mu = \frac{4\pi^2 a^3}{\tau^2}. \tag{6}$$

But, if (iii) be true, a^3/τ^2 has the same value for all planets. This implies that the proportionality factor μ is also the same for all (*Principia:* Book I, Prop. XV).

For orbits treated approximately as circular, and not elliptical, this result is apparent from (i) and (iii), much more simply. This was doubtless known to certain contemporaries of Newton, and his original 1666 treatment seems also to have been restricted to this idealized situation. All that is needed is the elementary formula for centripetal acceleration in uniform circular motion: $f = 4\pi^2 a/\tau^2$ (known, for example, to Huygens in 1659; cf. *Principia:* Scholium to Prop. IV). By (i) this holds for any planet, and by (iii) $\mu = fa^2 = 4\pi^2 a^3/\tau^2$ is the same for all.

Newton was thereby led to assign to the sun this value of the quantity μ. Similarly, by measuring a^3/τ^2 for each of Jupiter's moons and noting the near-uniformity of the results, a value of μ

could be assigned to Jupiter, and yet another to Saturn from observations of its satellites (*Principia:* Book III, Phenomena I and II).

1.5 Newtonian Postulate

By inductive generalization for all celestial bodies Newton (probably about 1679) adopted a standpoint amounting to the following formalization (e.g. *Principia:* Book III, General Scholium, or §26 of *The System of the World*).

> **Gravitational postulate for mass-points:** Any celestial object P has associated with it a positive scalar magnitude μ_P, not depending on position, motion, or time, called its gravitational mass, with dimensions of [length]^3/[time]^2. Any other celestial object Q has an acceleration μ_P/r^2 towards P, where r is the distance between P and Q, additionally to and independently of whatever acceleration Q would have in the absence of P.

In amplification, note that the imagined situations with and without P are compared on the understanding that the instantaneous positions and velocities of all other bodies are the same. Second, the stipulated independence implies that the resultant additional acceleration of Q, attributable to the presence of several bodies such as P, is to be found by *vector addition* of their separate contributions. Third, the postulate as stated suffices only to predict the motions of celestial bodies to within the approximation involved in neglecting their extensions in space, since the distance PQ is not precisely defined. A comprehensive statement valid also for continuous distributions of matter is deferred until later (§1.13). Fourth, the frame of reference is not particularized since *additional acceleration*, as distinct from total acceleration, is judged the same by all observers, whether using a sidereal frame or not (§2.1). This simple, but profound, invariance theorem in kinematics is indispensable to an understanding of mechanics, yet is apparently never mentioned.

To the above all-embracing postulate must be adjoined an extraneous 'working approximation', implicitly adopted though not directly stated by Newton. This concerns the field of acceleration attributable to matter that is outside the solar system. Without present evidence to the contrary, and having regard to the relatively minute size of the solar system (diameter of order 10^{-4} of the smallest stellar distances), it is assumed that the net acceleration field due to external matter is effectively uniform. More precisely, relative to a sidereal frame, all bodies in the solar system have at any moment the *same* (unspecified) accelerations in addition to their mutual inverse-square accelerations. Similarly, comparatively small and isolated star-systems elsewhere in the galaxy are assumed to move in a uniform field.

In appraising Newton's approach, the raison d'être of any scientific theory should be kept in mind. To begin with, no theory is ever unequivocally determined by observations, however extensive, nor arrived at by the exercise of logic alone. Many theories can always be constructed, and actually often are, to fit the same data within the presumed limits of experimental error. Conversely, any single substantiated datum suffices to delimit the range of validity of a particular theory, or even to dispose of it altogether. On the other hand, a discrepancy between experiment and theory does not necessarily imply a defect in the latter; in illustration we shall presently see that Newton's postulate is consistent only with a slightly modified version of Kepler's law (iii), which is indeed required by later and more accurate data. A theory remains in currency so long as it corresponds faithfully, within an allowed tolerance, to all observations in its proper domain; and especially if it delivers reliable predictions and prompts further investigations. In all these respects Newtonian mechanics is the classic example of a successful mathematical theory.†

1.6 Newtonian Frame: First Law of Motion

In analysing the motions of celestial bodies (natural or artificial) in the solar system we shall, for the present, continue to treat them as mass-points and ignore their extensions in space. This idealization is made for simplicity only, and does not affect the essence of the argument or the main conclusions (as will be verified in §1.13).

† For a succinct modern view of the philosophy of physical theories see R. von Mises: *Positivism*, Chapter 12 (George Braziller, New York 1956).

The additional acceleration, $\mathbf{a}_P$ say, of a body P by the rest of the solar system is then

$$\mathbf{a}_P = \sum_{Q \neq P} \{\mu_Q(\mathbf{r}_Q - \mathbf{r}_P)/|\mathbf{r}_Q - \mathbf{r}_P|^3\} \tag{1}$$

where the summation extends over all other bodies, typically Q. The origin, C say, of the position vectors $\mathbf{r}_P$, $\mathbf{r}_Q$, . . ., is arbitrary. By multiplying (1) by μ_P and adding all such equations vectorially, we obtain the overall property

$$\Sigma(\mu\mathbf{a})_P = \mathbf{0}. \tag{2}$$

The double summation on the right has vanished identically since it contains two equal and opposite terms in the product $\mu_P\mu_Q$ for each typical pair P and Q.

Now choose any sidereal frame together with an associated origin O in position $\mathbf{r}_O$ with respect to C. According to §1.5 the apparent acceleration of P, as seen by an observer in this frame, is

$$\ddot{\mathbf{r}}_P - \ddot{\mathbf{r}}_O = \mathbf{a}_P - \boldsymbol{\alpha}_O \tag{3}$$

where $\boldsymbol{\alpha}_O(t)$ is the same for all bodies but depends on the chosen frame and remains to be determined. For this purpose let C be specified as the **gravitational mass-centre** of the solar system at all times: that is,

$$\Sigma(\mu\mathbf{r})_P = \mathbf{0} \tag{4}$$

by analogy with the familiar definition of a centroid in geometry. Because of the sun's preponderance C is never far from the sun itself (*Principia:* Book III, Prop. XII). By multiplying each equation such as (3) by μ_P, and summing over the system, there follows $\boldsymbol{\alpha}_O = \ddot{\mathbf{r}}_O$ with the help of (2) and (4) twice differentiated. $\boldsymbol{\alpha}_O$ is thereby identified as the acceleration of O (or any other point fixed in the chosen frame) with respect to C. Equally, $-\boldsymbol{\alpha}_O$ is the acceleration of C in this frame, which means that the mass-centre moves as if the system were a mass-point in the external field.

From (1) and (3) we now have

$$\ddot{\mathbf{r}}_P = \sum_{Q \neq P} \{\mu_Q(\mathbf{r}_Q - \mathbf{r}_P)/|\mathbf{r}_Q - \mathbf{r}_P|^3\}, \tag{5}$$

for the acceleration of P with respect to C. This is the equation of motion in any sidereal frame relative to which C is at rest or moves with uniform velocity; such special frames are described as **Newtonian**. We may say, therefore, that the acceleration of any celestial body, observed from a Newtonian frame, is at all times *entirely* accounted for by the inverse-square contributions of other matter in the solar system.

In particular, the motion is unaccelerated in a Newtonian frame when, and only when, these contributions have zero resultant. This statement may be regarded as a modernized version of the **First Law of Motion,**† in a form investing it with real substance. By re-tracing the argument, the statement can be seen as tantamount to the 'working approximation' that was adjoined to the gravitational postulate. Newton's own version has been variously construed: to some it has simply seemed an axiomatic property of 'absolute space'; to others, a preliminary definition of 'force'; and to others again, a mere footnote to his Second Law.

We return to (5) in order to note a convenient equivalent. For this purpose a single scalar function of all the masses and mutual distances is introduced, called the **gravitational potential energy** of the system:

$$V = -\sum_{\text{pairs}} \{\mu_Q \mu_R / |\mathbf{r}_Q - \mathbf{r}_R|\} \tag{6}$$

summed over all pairs of bodies, typically Q and R. Its variation in an infinitesimal change of the configuration is

$$\delta V = \sum_{\text{pairs}} \{\mu_Q \mu_R (\mathbf{r}_Q - \mathbf{r}_R)\delta(\mathbf{r}_Q - \mathbf{r}_R)/|\mathbf{r}_Q - \mathbf{r}_R|^3\},$$

using the elementary relation $\mathbf{a}\delta\mathbf{a} = |\mathbf{a}|\delta|\mathbf{a}|$ which follows from the identity $\mathbf{aa} \equiv |\mathbf{a}|^2$ for any vector $\mathbf{a}$ ($|\delta\mathbf{a}|$ and $\delta|\mathbf{a}|$ must be distinguished).†† This may be written shortly as

$$\left.\begin{aligned} \delta V &= \sum\{(\text{grad}_P V)\delta\mathbf{r}_P\} \\ \text{where} \qquad \text{grad}_P V &= \sum_{Q \neq P} \{\mu_P \mu_Q (\mathbf{r}_P - \mathbf{r}_Q)/|\mathbf{r}_P - \mathbf{r}_Q|^3\}. \end{aligned}\right\} \tag{7}$$

† *Principia:* Book I. "Every body continues in its state of rest, or of uniform motion in a right line, unless compelled to change that state by forces impressed upon it."

†† The scalar product of two vectors **a** and **b** is indicated throughout this book by juxtaposition: **ab.** The superfluous dot notation will not be used.

The last expression is the formal vector gradient of V with respect to the coordinates of P only, disregarding the actual dependence (4) between position vectors from G.

Comparison with (5) now gives

$$(\mu\ddot{\mathbf{r}})_P = -\operatorname{grad}_P V, \text{ for each } P, \tag{8}$$

as a compact form of the equations of motion. The whole set can be gathered together in a single formula:

$$\Sigma(\mu\ddot{\mathbf{r}}\delta\mathbf{r})_P + \delta V = 0 \text{ whenever } \Sigma(\mu\delta\mathbf{r})_P = \mathbf{0}. \tag{9}$$

For this is implied by (8), on multiplying by $\delta\mathbf{r}_P$ scalarly and summing; and conversely (9) implies all the separate equations (8) with the help of the connexion (4) and the identity $\Sigma(\operatorname{grad}_P V) \equiv \mathbf{0}$ (to obtain the equation for P choose arbitrary $\delta\mathbf{r}_Q = \delta\mathbf{r}_R = \ldots$).

By the theory of second-order differential equations such as (8) a knowledge of the positions and velocities of all bodies in the system, at some present instant, uniquely determines in principle their motions at all times in the future (and in the past), unless collisions occur. However, in relation to any actual system, this determinism is illusory: first, because the equations generally cannot be solved by a finite combination of known functions, and the computational errors in approximate solutions (by cut-off series or iterations) cannot be kept small during an arbitrary interval of time; second, because actual initial data can never be absolutely precise, and the long-term effects of such uncertainty will be cumulative when the particular system of equations is inherently unstable; and third, simply because the mathematical model neglects various features that may contribute significantly in the long run. Accordingly, in practice, the solution must be re-started at the end of a suitable interval with the help of observational data at that time. This interval may be several decades or centuries, depending on the system in question and the accuracy aimed at.

1.7 Motion of Two Gravitating Bodies

The prototype problem in celestial mechanics is that of the periodic relative motion of two bodies idealized as mass-points whose accelerations are affected equally, at least in first approximation, by other matter. It should be realized that the additional accelerations of each body due to external influences need not be small in themselves, but that their *difference* must be small compared with the additional accelerations produced by the bodies in each other.† Examples are a planet and its major satellite; the sun and any planet; and an isolated binary (interacting double stars). Then, exactly as in the transition from (1) to (5) in §1.6 we may show that the mass-centre (G say) of the pair moves like an isolated mass-point under the external field; and that, in a sidereal frame, the acceleration of each body relative to G is just that due to the presence of the other.

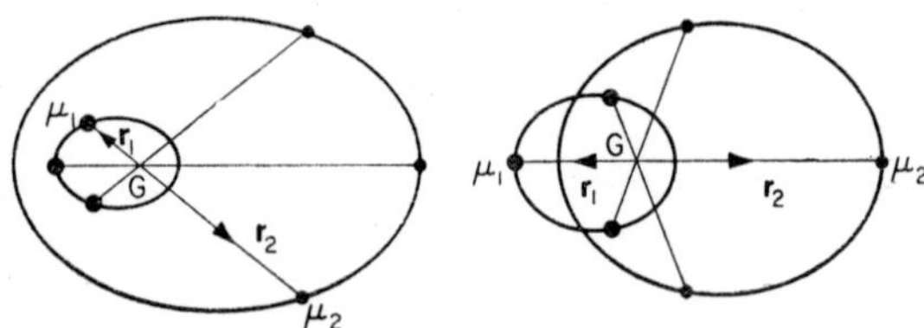

Fig. 1

By definition G is such that

$$\mu_1\mathbf{r}_1 + \mu_2\mathbf{r}_2 = \mathbf{0} \tag{1}$$

where $\mathbf{r}_1$ and $\mathbf{r}_2$ are the position vectors measured from G towards the bodies (Fig. 1 shows two of the possible configurations). Since the latter are necessarily always collinear with G, the orbits relative to G are similar, with linear dimensions in inverse ratio to the masses (*Principia:* Book I, Prop. LVII). The gravitational equations of motion, (5) or (8) of §1.6, reduce here to

$$\mu_1\ddot{\mathbf{r}}_1 = \mu_1\mu_2(\mathbf{r}_2 - \mathbf{r}_1)/|\mathbf{r}_2 - \mathbf{r}_1|^3 = -\mu_2\ddot{\mathbf{r}}_2, \tag{2}$$

only one being independent in view of (1). With its help they become

$$(\mu_1 + \mu_2)^2\ddot{\mathbf{r}}_1 = -\mu_2^3\mathbf{r}_1/r_1^3, \qquad (\mu_1 + \mu_2)^2\ddot{\mathbf{r}}_2 = -\mu_1^3\mathbf{r}_2/r_2^3, \tag{3}$$

both of which evidently define inverse-square central orbits with

† Thus, the acceleration of the moon by the earth is only about one half the accelerations of the moon and earth separately by the sun, but is about 100× their difference.

respect to G. To identify these more closely we note first that (3) implies

$$\frac{d}{dt}(\mathbf{r}_1 \wedge \dot{\mathbf{r}}_1) = 0 = \frac{d}{dt}(\mathbf{r}_2 \wedge \dot{\mathbf{r}}_2)$$

since vector products of parallel vectors vanish. Hence

$$\mathbf{r}_1 \wedge \dot{\mathbf{r}}_1 = 2\eta_1\mathbf{n}, \quad \mathbf{r}_2 \wedge \dot{\mathbf{r}}_2 = 2\eta_2\mathbf{n}, \quad \text{with } \eta_1/\eta_2 = (\mu_2/\mu_1)^2 \tag{4}$$

by (1), where $\mathbf{n}$ is some constant unit vector and η_1, η_2 are arbitrary scalar constants. From the direction of these vector products, the plane through G and the bound velocity vectors always has $\mathbf{n}$ as its normal, and so both orbits lie in a plane through G whose orientation is fixed in a sidereal frame. From the magnitudes of the vector products, the integration constants η_1 and η_2 are recognized as the rates of sweep of orbital areas about G.

When the orbits are ellipses, which we know from §1.4 to be possible under (3), the periods are necessarily identical (by the collinearity property) and the semi-major axes are in the ratio $a_1/a_2 = \mu_2/\mu_1$ (by the similarity property). The factors of proportionality in the inverse-square relations (3) are respectively equal to $4\pi^2a_1^3/\tau^2$ and $4\pi^2a_2^3/\tau^2$, as in (6) of §1.4. Solving these equations for the masses one obtains

$$\mu_1a_1 = \mu_2a_2 = a_1a_2\{2\pi(a_1 + a_2)/\tau\}^2. \tag{5}$$

Measurements of the period and mean distances from G thereby determine the separate masses. In passing, it may be mentioned that the ellipses intersect when their eccentricity exceeds

$$|\mu_1 - \mu_2|/(\mu_1 + \mu_2),$$

as may be shown by simple geometry.

Table 1 lists the values for the major planets, as multiples of the mass of the earth (without the moon). When no suitable satellite is available (as happens with Mercury and Venus, for example), the above method is inapplicable. The mass can then only be found indirectly, from the way in which the considered body perturbs the mutual orbit of two others; this involves the study of a 3-body system. As is wellknown, it was through the analysis of perturbation effects that Adams and Leverrier in 1846 independently inferred the existence and mass of Neptune from the seemingly anomalous behaviour of Uranus (whose orbit had not then been mapped in full, having been identified only in 1781); the anomalies were accounted for in still more detail by the discovery of Pluto in 1930.†

Values of the principal elements of the planetary orbits round the sun are also given in Table 1 for the present epoch. The sidereal

TABLE 1

Planet	Mass (earth units)	Sidereal period (years)	Semi-major axis (astronomical units)	Eccentricity	Inclination
Mercury	0·0543	0·2408	0·387	0·206	7° 0′
Venus	0·814	0·6152	0·723	0·0068	3° 24′
Earth	1	1	1	0·0167	0
Mars	0·107	1·8809	1·524	0·0934	1° 51′
Jupiter	317·4	11·865	5·202	0·0486	1° 18′
Saturn	95·0	29·456	9·538	0·056	2° 29′
Uranus	14·5	84·02	19·185	0·047	0° 46′
Neptune	17·6	164·8	30·06	0·0086	1° 46′
Pluto	~0·1	248·4	39·52	0·249	17° 10′

(Mass of sun)/(mass of earth) = 332,500

period is the quantity determinable with most precision (except that the values for Neptune and Pluto are provisional extrapolations from their partial orbits observed so far). Needless to say, no actual orbit is a perfect ellipse, or even a closed curve, owing to perturbations. To this extent the geometrical data relate to nominal 'mean ellipses'; thus, the semi-major axes are calculated from law (iii) itself, in the form $a^3/\tau^2 = 1$ where a is in astronomical units and τ is in years. Except in two cases the orbits deviate little from circles: whereas the difference between the apsidal distances is the small fraction $2e$ of their mean (e being the eccentricity), the fractional difference between major and minor axes is much smaller still, $\frac{1}{2}e^2$ very nearly. All the orbits are traversed in the same direction and their planes are fairly close; their inclinations to the plane of the ecliptic, all in the same sense, are listed in the final column.

† For an authoritative account see W. M. Smart: *John Couch Adams and the Discovery of Neptune* (Royal Astronomical Society 1947). An interesting commentary on Adams' method can be found in J. E. Littlewood: *A Mathematician's Miscellany*, pp. 117–32 (Methuen 1953). A recent study has been made by M. Grosser: *The Discovery of Neptune* (Oxford University Press 1962).

Several thousands of minor planets or asteroids are also known, mostly with orbits lying between those of Mars and Jupiter, the inclinations being widely spaced about a mean of 10°. The largest, Ceres, is estimated from its size and physical properties to have a mass of order $10^{-4} \times$ earth's mass. Finally, there are large numbers of comets whose individual masses are utterly negligible and whose orbits generally extend far outside the solar system. The majority of computed cometary orbits are near-parabolas, at least in the comparatively small arc accessible to observation, while the remainder are ellipses of large eccentricity. But perturbations, especially by Jupiter, prevent any definite periodicity and very few actual returns have ever been verified. One of the most regular comets is Halley's, for which $\tau \sim 76\frac{1}{2}$ years and $e = 0{\cdot}967$; the nominal value of its semi-major axis calculated from $a^3/\tau^2 = 1$ is

$$a = (76{\cdot}5)^{2/3} \sim 18{\cdot}0 \text{ a.u.}$$

By applying (5), or refinements of it, to the earth–moon system it is found that the mass-ratio is 0·01231 or 1/81·25 (average of several recent determinations). The earth's motion relative to G is detectable as a very small oscillation in the apparent motions of nearby celestial bodies seen from the earth, with a period equal to the lunar month. Due to many perturbation effects the orbit can only be treated as an ellipse in first approximation, with mean period 27·3217 days, eccentricity 0·055, and inclination 5° 8′; the average† earth–moon distance is 384,405 km. On this basis the semi-major axis of the orbit of the earth's centre about G is calculable as 384,405/82·25 = 4674 km; since the earth's mean radius is 6371 km G lies well inside its surface.

Kepler's law (iii) concerns the orbit of one body relative to another, not to their mass-centre. This mutual orbit is of course geometrically similar to the separate orbits about G; in particular, the respective rates of sweep of area are as the squares of the linear dimensions:

$$\eta/(\mu_1 + \mu_2)^2 = \eta_1/\mu_2^2 = \eta_2/\mu_1^2.$$

When the orbit is an ellipse its semi-major axis is $a = a_1 + a_2$

† The semi-major axis of an ellipse is not the *time*-average of the focal distance, but its average with respect to the auxiliary angle of the ellipse. That is, $a = \int_0^{\pi} r d\phi/\pi$ where $r = a(1 - e\cos\phi)$. In astronomy the term 'average distance' usually refers simply to the mean of the least and greatest apsidal distances.

(Fig. 1) and the period is again τ. The connexion between these quantities follows directly from (5):

$$\frac{4\pi^2 a^3}{\tau^2} = \mu_1 + \mu_2. \tag{6}$$

With μ_1 as the mass of the sun and μ_2 that of a planet with all its satellites, this is the proper interpretation of μ in (6) of §1.4 and law (iii) must be modified accordingly (*Principia:* Book I, Prop. LIX). a^3/τ^2 is not strictly the same for all the planets, but its fractional variation is less than 10^{-3}; when (6) is used to re-calculate the values of a in Table 1 to the same number of decimal places, only the entry for Jupiter changes (to 5·203).

The equation of motion in the relative orbit is obtainable as the appropriate combination of equations (2):

$$\ddot{\mathbf{r}} = -\mu\mathbf{r}/r^3 \quad \text{with} \quad \mu = \mu_1 + \mu_2, \tag{7}$$

where $\mathbf{r}$ is now written for $\mathbf{r}_1 - \mathbf{r}_2$. The 'integral of area' is

$$\mathbf{r} \wedge \dot{\mathbf{r}} = 2\eta\mathbf{n}, \tag{8}$$

while by (3) of §1.4 the speed–distance dependence is

$$\frac{v^2}{\mu} - \frac{2}{r} = \text{constant}, \quad -c \text{ say}. \tag{9}$$

With $v = 2\eta/p$ this becomes

$$\frac{l}{p^2} = \frac{2}{r} - c \quad \text{where} \quad l = 4\eta^2/\mu. \tag{10}$$

In particular, the positive root(s) of the quadratic

$$\left(\frac{l}{d}\right)^2 - 2\left(\frac{l}{d}\right) + lc = 0$$

furnish the apsidal distance(s) d in the positions where $p = r$. If d_1 and d_2 are the actual roots, we can plainly write

$$\frac{l}{d_1} = 1 + e, \quad \frac{l}{d_2} = 1 - e, \quad \text{where } e \geqslant 0, \tag{11}$$

and then

$$c = \frac{l}{d_1 d_2} = \frac{1 - e^2}{l} = \frac{1 - e}{d_1} = \frac{1 + e}{d_2} \tag{12}$$

with a single apse if and only if $c \leqslant 0$ and $e \geqslant 1$. In fact, (10) is known to be the pedal equation of a conic with one focus at $r = 0$, semi-latus rectum l and eccentricity e. When $c > 0$ it is an ellipse and when $c < 0$ it is the near-branch of a hyperbola, respectively with semi-major and semi-transverse axis $a = 1/|c|$. It is a parabola when $c = 0$, and one sees from (9) that the speed at any distance is then always $\sqrt{2} \times$ the speed in a circular orbit of that radius; by this property Newton showed that certain comets followed parabolas, within observational error (*The System of the World:* §73).

1.8 Three Gravitating Bodies: Lagrange Configurations

In the 2-body problem it is obvious that the orbits about the mass-centre G could in particular be concentric circles of radii a_1 and a_2, with the bodies at the fixed distance $a = a_1 + a_2$. Their join rotates in a sidereal frame at rate $\omega = 2\pi/\tau$ where τ is given by (6) of §1.7; that is,

$$\omega^2 = (\mu_1 + \mu_2)/a^3. \tag{1}$$

Conversely, if in this frame the bodies are, at some instant, distances a_1 and a_2 from G and have speeds $a_1\omega$ and $a_2\omega$ relative to G and transversely to their join, they will thenceforward describe circular orbits. Any other 'initial' conditions are followed by general conics.

It is natural to enquire whether 3 mass-points can move similarly in concentric orbits, maintaining some unvarying configuration rotating at a suitable rate Ω in a sidereal frame (Lagrange 1772). We examine first the possible collinear configurations. To fix ideas, regard as given two of the masses μ_1 and μ_2 together with their separation a. Let λa be the coordinate of the third mass μ with respect to the first, taken in the sense shown in Fig. 2. Then λ and Ω are the unknowns to be determined. The coordinate of the mass-centre G with respect to the first body is γa, where

$$\gamma = (\lambda\mu + \mu_2)/(\mu_1 + \mu_2 + \mu).$$

From this the centripetal acceleration of each body relative to G in the assumed motion can be obtained, and then identified with the algebraic sum of the inverse-square accelerations of this body due to the other two. From μ_1 and μ_2 in turn one finds that

$$\frac{a^3\Omega^2}{\mu_1 + \mu_2 + \mu} = \frac{\dfrac{\lambda\mu}{|\lambda|^3} + \mu_2}{\lambda\mu + \mu_2} = \frac{\dfrac{(\lambda - 1)\mu}{|\lambda - 1|^3} - \mu_1}{(\lambda - 1)\mu - \mu_1}. \tag{2}$$

The equation for μ is just a combination of these and so adds nothing new. As a check on (2) it may be verified that when the masses are all equal the expected values $\lambda = -1, \frac{1}{2}, 2$ are solutions; the corresponding configurations differ only in scale, with one body midway between the others. For arbitrary masses the quintic equation for λ has to be solved by an iterative numerical procedure. There are three real roots, for each of which the interior body is different.

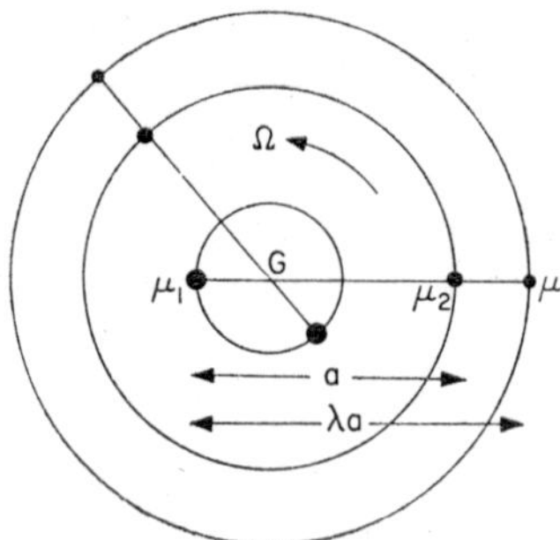

FIG. 2

When the third mass is vanishingly small, the spin is effectively as in (1) while λ satisfies

$$\mu_1\lambda\left(\frac{1}{|\lambda|^3}-1\right)+\mu_2(\lambda-1)\left(\frac{1}{|\lambda-1|^3}-1\right)=0. \tag{3}$$

For example, when the first and second bodies are the earth and moon, two of the Lagrange positions are near the moon (roughly 57,500 km on either side). There is a suggestion that despite perturbation effects a lunar communications system could be provided by artificial satellites in or near these stations.

Consider, next, the possibility of a triangular configuration having constant spin Ω in a plane of fixed orientation in a sidereal frame (Fig. 3). Relative to G the accelerations are $-\mathbf{r}_i\Omega^2$ ($i = 1, 2, 3$). It is apparent that the resultant gravitational acceleration of each body has this required centripetal property only if the triangle is equilateral. In that event the accelerations produced in, say, the first body have a resultant

$$\frac{\mu_2}{a^3}(\mathbf{r}_2-\mathbf{r}_1)+\frac{\mu_3}{a^3}(\mathbf{r}_3-\mathbf{r}_1) = -\frac{(\mu_1+\mu_2+\mu_3)}{a^3}\mathbf{r}_1$$

where a is the length of side, since

$$\mu_1\mathbf{r}_1 + \mu_2\mathbf{r}_2 + \mu_3\mathbf{r}_3 = 0$$

by the definition of G. Thus, the assumed motion persists if

$$a^3\Omega^2 = \mu_1 + \mu_2 + \mu_3. \tag{4}$$

Analogous results have been obtained for plane systems where $n > 3$.

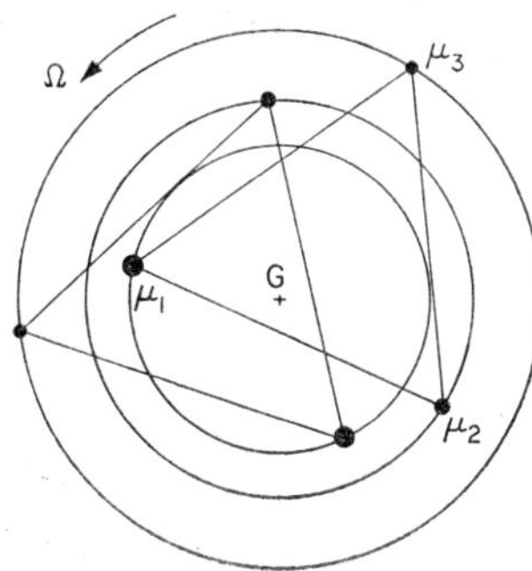

FIG. 3

It appears that the Trojan group of asteroids form with the sun and Jupiter a near-equilateral triangle, apart from perturbation oscillations. The two triangular positions with respect to the earth and moon, together with the far collinear position, have been proposed for possible occupation by space observatories.

1.9 The Restricted Problem of 3 Bodies

This is the situation where one mass μ is relatively so small that it has negligible effect on the other two, μ_1 and μ_2, and where the common plane maintains a constant orientation in a sidereal frame. Additionally, the large masses are supposed to remain a fixed distance apart, a say; their join then has a uniform spin ω, given by (1) in §1.8.

It is simplest to specify the position P of the small body by coordinates (ξ, η) on rectangular axes rotating at rate ω, with origin at the mass-centre G (Fig. 4). Then the acceleration of P relative to G, and in a sidereal frame, has components

$$(\ddot{\xi} - 2\omega\dot{\eta} - \omega^2\xi, \quad \ddot{\eta} + 2\omega\dot{\xi} - \omega^2\eta)$$

on the axes of reference. These expressions are derived most conveniently as the real and imaginary parts of

$$\frac{d^2}{dt^2}\{(\xi + i\eta)e^{i\omega t}\} \quad \text{evaluated at } t = 0.$$

Since the apparent acceleration of P relative to the rotating frame of reference is just $(\ddot{\xi}, \ddot{\eta})$, this result illustrates how accelerations in

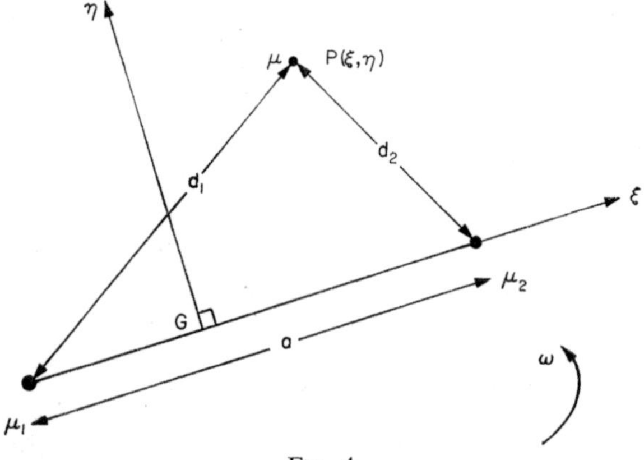

FIG. 4

mutually rotating frames differ by terms involving both position and velocity; the general connexion is obtained later in §2.1.

By (6) in §1.6 the gravitational potential energy of the system is

$$V(\xi, \eta) = -\left(\frac{\mu\mu_1}{d_1} + \frac{\mu\mu_2}{d_2} + \frac{\mu_1\mu_2}{a}\right)$$

where distances d_1 and d_2 are calculable functions of (ξ, η). Then, by (8) in §1.6, the equation of motion of P has components

$$\ddot{\xi} - 2\omega\dot{\eta} = \partial U/\partial\xi, \qquad \ddot{\eta} + 2\omega\dot{\xi} = \partial U/\partial\eta, \tag{1}$$

where

$$U(\xi, \eta) = \tfrac{1}{2}\omega^2(\xi^2 + \eta^2) + \frac{\mu_1}{d_1} + \frac{\mu_2}{d_2}, \quad \text{with } a^3\omega^2 = \mu_1 + \mu_2. \tag{2}$$

Possible positions of relative rest, the so-called 'points of libration', are the solutions of $\partial U/\partial \xi = 0 = \partial U/\partial \eta$, and are just the five Lagrange positions of §1.8 for the present special system. Equations (1) and (2) initiated the extensive modern theory of the motion of the moon in the field of the earth and sun, and of asteroids in the field of the sun and a major planet.

The solution family of orbits of P is not obtainable in closed form for all time, and the analysis has been carried farther only in regard to particular aspects of the problem and by means of series expansions. It is known, for example, that periodic closed orbits exist around each massive body and also around each point of libration, within certain neighbourhoods. However, the collinear positions are definitely unstable in the sense that there are certain types of vanishingly-small disturbances that cause a body resting there to depart by a finite distance; the manner and rapidity of departure has not been thoroughly investigated. The triangular positions also are unstable when $|\mu_1 - \mu_2|/(\mu_1 + \mu_2) < \sqrt{(23/27)}$, which happens when the smaller of the two masses exceeds 0·03852... of the larger; with a more extreme mass-ratio the positions are stable in first approximation, but whether rigorously so is not known.†

Useful partial information can be gained from a certain 'first integral' of (1), introduced by Jacobi (1836). To obtain it multiply the first equation by $\dot{\xi}$, the second by $\dot{\eta}$, add, and integrate with respect to time. The result is

$$\tfrac{1}{2}(\dot{\xi}^2 + \dot{\eta}^2) = U(\xi, \eta) - c \tag{3}$$

where c is an arbitrary constant, determined by initial data for the position and speed of P. Although its actual path cannot be calculated from (3) alone, it nevertheless at least makes plain that P cannot go where the local U-values are less than the particular orbital c-value. That is, the path certainly lies within the region

$$U(\xi, \eta) \geqslant c \tag{4}$$

† For more details and references see E. T. Whittaker: *A Treatise on the Analytical Dynamics of Particles and Rigid Bodies*, §181 *et seq.* (Cambridge University Press 1937, 4th Edition). Also A. Wintner: *The Analytical Foundations of Celestial Mechanics*, Chap. VI (Princeton University Press 1941) for a comprehensive and advanced discussion; and E. Finlay-Freundlich: *Celestial Mechanics* (Pergamon Press 1958) or F. R. Moulton: *An Introduction to Celestial Mechanics* (Macmillan 1914, 2nd revised edition) or S. W. McCuskey: *Introduction to Celestial Mechanics* (Addison-Wesley 1963) for more elementary accounts.

of the moving frame, and furthermore that, if this region consists of more than one disjunct part, the path is confined to the part where it started. Thus (3) supplies information of a negative, or exclusive, kind when the level contours of the function U are known.

Their broad features can be established fairly easily. Note first from (2) that U is essentially positive and that it is unbounded at infinity and near the two bodies. The Lagrange positions are its

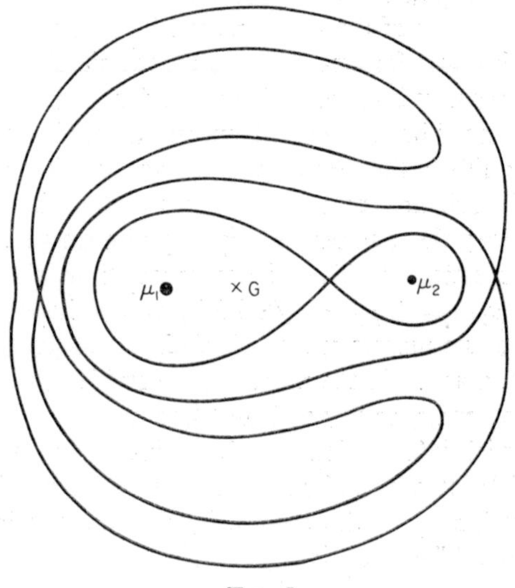

FIG. 5

only stationary values, as already remarked. It is straightforward to show that the triangular positions are absolute minima, while the collinear ones are merely 'cols' or 'saddle-points'. Through each of the latter, therefore, passes a certain contour with a double point at the col; the purely schematic Fig. 5 indicates the appearance of these particular contours when $\mu_1 > \mu_2$, and hence qualitatively the rise and fall of the whole U-surface.

To illustrate the kind of deduction such a diagram makes possible, suppose that the small mass is a space vehicle launched from the first body towards the second. When the initial rocket-thrust ceases, let the vehicle have speed u_0 in the moving frame and be somewhere on the contour c_0; then $c = c_0 - \frac{1}{2}u_0^2$ for the ensuing free-flight orbit. If, now, c were greater than the value, c_i say, corresponding to the contour through the inner Lagrange point, the vehicle could never penetrate near its target. A minimum requirement for success is therefore that u_0 should exceed $\sqrt{2}(c_0 - c_i)^{\frac{1}{2}}$. Prediction of the orbits of actual space vehicles is a far more complex matter, since many disturbing factors must be incorporated and high accuracy is called for.

1.10 The n-Body Problem: First Integrals

The complexities of the general problem of 3 bodies, the orientation of whose plane varies, can by now readily be imagined. We turn, therefore, from considering the separate orbits in detail and examine instead some simple properties of the motions collectively. For this purpose it is just as easy to deal with an arbitrary number n of free bodies. It is, as always, supposed that this system (which may in particular be the entire solar system) is isolated in the sense that each body experiences the same additional acceleration due to matter outside the system. One overall property is already known (cf. §1.6): the mass-centre G moves with this acceleration in the adopted sidereal frame.

With regard to the motions relative to G, the gravitational potential energy V immediately suggests itself as an overall quantity whose variation in time might profitably be investigated. From (9) in §1.6, by taking for $\delta\mathbf{r}_P$ the positional change $d\mathbf{r}_P$ in time dt in an actual motion, one has

$$\frac{dV}{dt} = -\Sigma(\mu\dot{\mathbf{r}}\ddot{\mathbf{r}})_P.$$

This integrates to

$$T + V = \text{constant}, \quad E \text{ say}, \tag{1}$$

where $$T = \tfrac{1}{2}\Sigma(\mu\dot{\mathbf{r}}^2)_P \quad \text{and} \quad V = -\sum_{\text{pairs}} \frac{\mu_Q\mu_R}{|\mathbf{r}_Q - \mathbf{r}_R|}. \tag{2}$$

The scalar quantity T is called the **kinetic energy** (of gravitational mass) in the motion relative to G. It is essentially positive whereas V is essentially negative; their constant sum can have either sign, and we shall see later that the long-term behaviour of the system depends critically on it. By the definition of mass-centre, (4) of

§1.6, T and V can be expressed entirely in terms of the positions and speeds of any $(n - 1)$ of the bodies relative to G; for example, when $n = 2$, there are two such equivalent forms of (1) and these are just the speed–distance relations for either orbit about G, corresponding to the integrals of equations (3) in §1.7.

Another expression for T, not implicitly involving G, but directly in terms of the mutual relative velocities of the bodies themselves, is

$$T = \tfrac{1}{2}\sum_{\text{pairs}}\{\mu_Q\mu_R|\dot{\mathbf{r}}_Q - \dot{\mathbf{r}}_R|^2\}/\Sigma\mu_P. \tag{3}$$

For the difference (2) − (3) is identically equal to $\tfrac{1}{2}\{\Sigma(\mu\dot{\mathbf{r}})_P\}^2/\Sigma\mu_P$, which vanishes by the definition of G.

Returning to the general equations of motion, (5) in §1.6, we form next the vector product of position and mass-acceleration for each body:

$$(\mathbf{r} \wedge \mu\ddot{\mathbf{r}})_P = \sum_{Q\neq P}\{\mu_P\mu_Q\mathbf{r}_P \wedge \mathbf{r}_Q/|\mathbf{r}_P - \mathbf{r}_Q|^3\}.$$

Hence

$$\frac{d}{dt}\Sigma(\mathbf{r} \wedge \mu\dot{\mathbf{r}})_P = \Sigma(\mathbf{r} \wedge \mu\ddot{\mathbf{r}})_P = \mathbf{0}$$

since all $(\dot{\mathbf{r}} \wedge \dot{\mathbf{r}})_P = \mathbf{0}$ and the terms on the right cancel in pairs in the double summation. Thus,

$$\Sigma(\mathbf{r} \wedge \mu\dot{\mathbf{r}})_P = \text{constant vector, } \mathbf{h} \text{ say.} \tag{4}$$

h is the **moment of momentum** (or angular momentum) about G of the motion relative to G. According to (4) its magnitude and direction are constant in time for an isolated gravitating system. The plane through G perpendicular to **h** maintains a constant orientation in a sidereal frame, and was called by Laplace the **invariable plane** of the system.

For the mass-point idealization of the solar system, in particular, it is also referred to as the **astronomical plane.** Its situation is such that the orbital planes of Jupiter and Saturn lie on either side, while the present inclination of the ecliptic is 1° 35′. This inclination slowly oscillates under planetary perturbations.

Finally, the sum of the scalar products of position and mass-acceleration invites investigation. From the equations of motion in the form (8) in §1.6:

$$\Sigma(\mu\mathbf{r}\ddot{\mathbf{r}})_P = -\Sigma(\mathbf{r}_P\,\text{grad}_P V).$$

Now V is homogeneous of degree -1 in the components of the position vectors, and so the preceding right side is identically equal to V by Euler's theorem for homogeneous functions (or by direct verification from (7) of §1.6). Whence

$$\Sigma(\mu \mathbf{r}\ddot{\mathbf{r}})_P = V. \tag{5}$$

This is not of much interest in itself but is needed in the developments of §1.11.

The energy principle (1), with the three components of the moment of momentum principle (4), together constitute **first integrals** of the original equations of motion relative to G, being differential relations one less in order (originally due to Euler about 1744–8). Their distinctive characteristic is that they involve the velocities in *algebraic* form, the same for all times and arbitrary initial conditions. It was proved by Bruns (1887) for $n = 3$, and by Painlevé (1898) for $n > 3$, that other integrals either do not have such properties or are merely dependent combinations of these four.† Jacobi's integral in the restricted problem of 3 bodies is actually just such a combination.

When $n = 2$ the above first integrals are in themselves a complete replacement of the equations of motion. Thus, equation (1) with (3) is the speed–distance relation for the relative orbit:

$$\tfrac{1}{2}v^2 - \frac{\mu_1 + \mu_2}{r} = \frac{\mu_1 + \mu_2}{\mu_1\mu_2} E \tag{6}$$

where r is the mutual distance and v the relative speed. And (4) is reducible to any of the equivalent forms (4) and (8) in §1.7, with

$$\mathbf{h} = 2(\mu_1\eta_1 + \mu_2\eta_2)\mathbf{n} = \frac{2\mu_1\mu_2}{\mu_1 + \mu_2}\eta\mathbf{n}. \tag{7}$$

Comparison of (6) here with (9) in §1.7 identifies E with $-\frac{1}{2}\mu_1\mu_2 c$ where $c = 1/a$ for an ellipse, 0 for a parabola, and $-1/a$ for the

† When $n = 2$ another independent algebraic integral exists, first noticed by Hamilton. Since $\mu\dot{\theta} = 2\mu\eta/r^2 = 2\eta f$,

$$\frac{d}{dt}\left(\frac{\mu\mathbf{r}}{r} + 2\eta\mathbf{n} \wedge \mathbf{v}\right) = \mathbf{0}.$$

The vector constant of integration has magnitude μe and is directed from the focus to the intersection of the orbital normal with the major axis.

near-branch of a hyperbola. In short, the total energy suffices by itself to determine the semi-major or semi-transverse axis, and hence also the period when the orbit is an ellipse:

$$\left(\frac{\tau}{2\pi}\right)^2 = \left(\frac{\mu_1\mu_2}{-2E}\right)^3 \Big/ (\mu_1 + \mu_2). \tag{8}$$

The moment of momentum, on the other hand, determines the semi-latus rectum:

$$l = \frac{\mu_1 + \mu_2}{(\mu_1\mu_2)^2}\, h^2 \tag{9}$$

from (10) in §1.7 and (7) above. Finally, the eccentricity is such that

$$e^2 = 1 + \frac{2(\mu_1 + \mu_2)}{(\mu_1\mu_2)^3}\, h^2 E \tag{10}$$

since $e^2 = 1 - cl$ by (12) in §1.7.

1.11 Dispersion of an *n*-Body System

The first integrals do not, as they stand, tell how the overall configuration of the system changes in the course of time: more must be extracted from the original equations of motion. The scalar quantity

$$J = \Sigma(\mu \mathbf{r}^2)_P \tag{1}$$

suggests itself as a global measure of the dispersion of mass away from the mass-centre. J is called the **polar moment of inertia** about G.† An equivalent expression in terms of mutual distances only is

$$J = \sum_{\text{pairs}} \{\mu_Q \mu_R |\mathbf{r}_Q - \mathbf{r}_R|^2\} / \Sigma \mu_P \tag{2}$$

in analogy to (3) of §1.10.

† The analogous quantity in gas dynamics of a molecular system is more usually called the 'virial of Clausius'.

It is easy to show that the polar moment of inertia about any other point O exceeds J by the polar moment about O of the total mass concentrated at G. The mass-centre could therefore be characterized as the point with least polar moment of inertia.

Double differentiation of (1) produces recognizable groups of terms:

$$\tfrac{1}{2}\ddot{J} = \Sigma(\mu\dot{\mathbf{r}}^2)_P + \Sigma(\mu\mathbf{r}\ddot{\mathbf{r}})_P.$$

Thus, from (1), (2) and (5) of §1.10,

$$\tfrac{1}{2}\ddot{J} = 2T + V \quad \text{where } T + V = E. \tag{3}$$

This result is due to Lagrange, and independently also to Jacobi who made it the basis of the following investigation.

Suppose that E is positive. Then $\ddot{J}$ has a positive lower bound for all times, since $\tfrac{1}{2}\ddot{J} = 2E - V > 2E$ as V is always negative. $J(t)$ is therefore a strictly convex function (which is to say that its graph has a curvature of the same sign everywhere) and moreover has no asymptotes as $t \to \pm\infty$ (which would imply vanishingly small $\ddot{J}$). It follows that J must be unbounded both in the future and in the past. We conclude from (2) that the greatest mutual distance is similarly unbounded, that infinite dispersion is inevitable for at least part of the system, and that no permanent 'capture' is possible. It cannot, however, be inferred that at any given instant in the future (or the past) the most distant body is eventually always the same (except trivially when $n = 2$).

Consider, next, the special case $E = 0$. If the greatest mutual distance were bounded above for all times, so also would be the potential energy: $V \leqslant -c$ say, where c is some positive constant. Then $\ddot{J} = -2V \geqslant 2c$, which implies as before that J is unbounded. The contradiction disproves the initial supposition and shows that infinite dispersion certainly occurs in part or whole as $t \to \pm\infty$.

Lastly, suppose that $E < 0$. Since $T \geqslant 0$, then $V \leqslant E < 0$, which immediately proves the impossibility of *total* dispersion, in the sense that all bodies recede indefinitely from each other (so that $V \to 0$). Specifically, the least mutual distance can never exceed $\sum_{\text{pairs}} (\mu_Q\mu_R)/(-E)$ as is apparent from the definition (6) in §1.6. On the other hand, when $n \geqslant 3$, all bodies *may* be indefinitely far from the mass-centre G. But nothing can be inferred from (3) about the certainty or otherwise of partial dispersion to infinity; however, by an entirely different method it can be shown that, if dispersion occurs, it does so both in the future and in the past.† On the other hand, if

† J. E. Littlewood: *A Mathematician's Miscellany*, pp. 133–4 (Methuen 1953). See also R. Kurth: *Introduction to Mechanics of Stellar Systems*, Chapter IV (Pergamon Press 1957).

J remains bounded as $t \to \pm\infty$ then so does $\dot{J}$, with the result that

$$\frac{1}{t}\int_0^t \ddot{J}\,dt = \frac{1}{t}\Big[\dot{J}\Big]_0^t \to 0 \quad \text{as} \quad t \to \pm\infty.$$

It follows from (3) that the long-term time-averages of the kinetic and potential energies satisfy

$$2T_{\text{ave}} = -V_{\text{ave}} = -2E > 0. \tag{4}$$

Mean-value properties such as this, in any dynamical context, are called 'virial theorems'.

Naturally, Jacobi's criterion may need modification in regard to actual systems, where other factors could become important over long periods of time; for example, collisions, dissipation of mechanical energy, and perturbations by distant matter.

It is instructive to see how the criterion operates when $n = 2$. Since partial dispersion is not then a possibility, the conclusions are precise: the distance between the two bodies is unbounded in the future and in the past when $E \geqslant 0$ and is bounded when $E < 0$. And this is confirmed by (6) of §1.10 which, with the help of the other first integral pv = constant, becomes the pedal equation of a hyperbola when $E > 0$, a parabola when $E = 0$, and an ellipse when $E < 0$ [cf. (10) of §1.7]. Notice carefully that, although (6) of §1.10 alone suffices to prove the orbit bounded when $E < 0$, it merely indicates that the orbit *may* be unbounded when $E \geqslant 0$ [in that (6) then formally admits real v as $r \to \infty$]. That the orbit in fact extends to infinity when $E \geqslant 0$ follows only from *full* use of the equations of motion, either explicitly after complete integration or implicitly through just such an argument as Jacobi's. (A similar point arose in §1.9 in connexion with the integral in the restricted problem of 3 bodies.)

1.12 Some Inequalities for an *n*-Body System

In the work of Jacobi, just discussed, no use is made of the conservation principle for the moment of momentum. With its help one would expect to find additional, sharper, theorems on the overall behaviour of the system. However, these have proved remarkably elusive and nothing significant was achieved until 1913 when

Sundman discovered a simple but deep result when $n = 3$.† A proof for arbitrary n is now presented here.

In order to explore what kind of theorem might reasonably be sought, consider first $n = 2$ and the identity

$$\mathbf{r}^2\mathbf{v}^2 \equiv (\mathbf{r}\mathbf{v})^2 + (\mathbf{r} \wedge \mathbf{v})^2$$

where $\mathbf{r}$ as usual is the mutual position vector and $\mathbf{v}$ is the relative velocity. Expressed in terms of generalizable quantities this is

$$2TR^2 \equiv (R\dot{R})^2 + h^2 \quad \text{where } R^2 = J \quad (n = 2) \tag{1}$$

since

$$\frac{2T}{\mathbf{v}^2} = \frac{h}{|\mathbf{r} \wedge \mathbf{v}|} = \frac{R^2}{r^2} = \frac{\mu_1\mu_2}{\mu_1 + \mu_2}$$

when $n = 2$, by (3) and (7) in §1.10 and (2) in §1.11. (It must be kept in mind that $\mathbf{r}\mathbf{v} = \mathbf{r}\dot{\mathbf{r}} = r\dot{r}$ but $\neq rv$ in general.) We now seek a similar relation for arbitrary n between the kinetic energy T, the magnitude h of the moment of momentum, and the weighted 'overall radius' R and its time-derivative. The relation turns out to be, not an equality, but an inequality.

We begin with the remark that

$$(\mathbf{a}^2 + \mathbf{b}^2 + \ldots)(\mathbf{l}^2 + \mathbf{m}^2 + \ldots) \geqslant (\mathbf{a}\mathbf{l} + \mathbf{b}\mathbf{m} + \ldots)^2 + (|\mathbf{a} \wedge \mathbf{l}| + |\mathbf{b} \wedge \mathbf{m}| + \ldots)^2 \tag{2}$$

where $\mathbf{a}, \mathbf{b}, \ldots$ and $\mathbf{l}, \mathbf{m}, \ldots$ are any vectors (not necessarily coplanar) and each bracket contains precisely the same number of terms. The resemblance to the previous simple identity is evident but the equality holds if and only if the magnitudes satisfy $a/l = b/m = \ldots$ and all the pairs of directions $\mathbf{a}$ and $\mathbf{l}$, $\mathbf{b}$ and $\mathbf{m}$, . . . include the same angle. Alternatively, (2) can be written entirely in scalars as

$$(a^2 + b^2 + \ldots)(l^2 + m^2 + \ldots) \geqslant (al\cos\theta + bm\cos\phi + \ldots)^2 + (al\sin\theta + bm\sin\phi + \ldots)^2 \tag{3}$$

where $\mathbf{a}$ and $\mathbf{l}$ include the angle θ, $\mathbf{b}$ and $\mathbf{m}$ the angle ϕ, etc. The

† K. F. Sundman: Mémoire sur le probléme des trois corps. *Acta Mathematica* 1913, **36,** 105–79. A proof for arbitrary n has not been traced in the literature, although said to be possible by G. D. Birkhoff: *Dynamical Systems,* Chap. IX (American Mathematical Society Colloquium Publications, Vol. IX, 1927).

proof by verification is immediate: the expansion of the left side of (3) is

$$(al)^2 + (bm)^2 + (am)^2 + (bl)^2 + \dots .$$

while the expansion of the right side is

$$(al)^2 + (bm)^2 + 2(am)(bl)\cos(\theta - \phi) + \dots .$$

The inequality is therewith plainly in the direction stated, with equality if and only if $am = bl$, $\theta = \phi$, etc., for all possible pairs. Alternatively, and with more insight, (2) can be regarded as the vector analogue of the complex form of Cauchy's inequality:

$$(\alpha\bar{\alpha} + \beta\bar{\beta} + \dots)(\lambda\bar{\lambda} + \mu\bar{\mu} + \dots) \\ \geqslant (\alpha\bar{\lambda} + \beta\bar{\mu} + \dots)(\bar{\alpha}\lambda + \bar{\beta}\mu + \dots) \quad (4)$$

where $\alpha, \beta, \dots$ and $\lambda, \mu, \dots$ are complex numbers, a bar indicates the conjugate, and the equality holds if and only if $\alpha/\lambda = \beta/\mu = \dots .$ This is a consequence of

$$(\alpha z - \lambda)(\bar{\alpha}\bar{z} - \bar{\lambda}) + (\beta z - \mu)(\bar{\beta}\bar{z} - \bar{\mu}) + \dots \geqslant 0 \text{ for any } z,$$

as may be proved by writing this as a bilinear form in $(z, \bar{z})$ and 'completing the square' (cf. generation of the discriminant of an ordinary quadratic). The connexion between (3) and (4) is established by putting a and l equal to the moduli of α and λ, and θ equal to the difference of their arguments. The right sides of (3) and (4) are, of course, just different ways of writing the squared modulus of a complex quantity, namely as the sum of squares of its real and imaginary parts or as the product of the quantity and its conjugate.

Returning to (2) we draw from it a weaker inequality suited to the present purpose:

$$(\mathbf{a}^2 + \mathbf{b}^2 + \dots)(\mathbf{l}^2 + \mathbf{m}^2 + \dots) \geqslant (\mathbf{al} + \mathbf{bm} + \dots)^2 \\ + (\mathbf{a} \wedge \mathbf{l} + \mathbf{b} \wedge \mathbf{m} + \dots)^2. \quad (5)$$

This follows obviously since the sum of the lengths of a set of vectors is not less than the length of their vector resultant. The equality holds if and only if the sets **a**, **b**, . . . and **l**, **m**, . . . are coplanar and differ simply by a rigid rotation and magnification. If, now, we choose for **a**, **b**, . . . the n weighted position-vectors $\sqrt{\mu_P}\mathbf{r}_P$ and for **l**, **m**, . . . the weighted velocities $\sqrt{\mu_P}\dot{\mathbf{r}}_P$, (5) supplies

$$2T \geqslant \dot{R}^2 + \frac{h^2}{R^2} \tag{6}$$

with equality (when $n \geqslant 3$) if and only if the system is coplanar and moves instantaneously in the plane so as to preserve geometric similarity (i.e., the relative velocities are proportional to the radii from G and rotated equally from them). Combining (6) with the Lagrange–Jacobi relation, (3) in §1.11, one has

$$2R\ddot{R} + \dot{R}^2 - \frac{h^2}{R^2} \geqslant 2E,$$

which is the same as

$$\frac{\dot{S}}{\dot{R}} \geqslant 0 \quad \text{where} \quad \frac{S}{R} = \dot{R}^2 + \frac{h^2}{R^2} - 2E. \tag{7}$$

This is the generalization of Sundman's basic result, to the effect that the function S perpetually tends to increase or decrease synchronously with R (or J).

This inequality was primarily applied by Sundman and Birkhoff to the study of conditions under which collisions may occur between two or more bodies when $n = 3$. We give here just one illustration of the type of argument, in showing the impossibility of an n-fold collision, even in a limiting sense, when $h \neq 0$. Supposing it were possible, so that J and $R \to 0$ from above (not necessarily monotonically) as $t \to t_0$ say, then $V \to -\infty$ and $\ddot{J} = 2(2E - V)$ is positive throughout some interval leading up to t_0. Hence, sufficiently near t_0, there is some interval in which J does not oscillate but decreases monotonically (with $\dot{J}$ remaining bounded). Therefore, by the inequality (7), S tends to decrease. On the other hand, by the definition (7), S is unbounded above as $R \to 0$ because of the term h^2/R ($h \neq 0$), the two others being bounded below. The contradiction shows that, contrary to the supposition, J and R each have a fixed least upper bound in all arbitrarily small intervals leading up to t_0, and consequently so has the largest mutual distance in the system.

1.13 Generalized Newtonian Postulate

In order to deal more adequately with actual bodies, than by idealizing them as mass-points, we need to amplify the basic gravitational postulate in §1.5 by re-phrasing it in terms of mass-density. For this purpose the standpoint customary in classical field-theories is adopted, the mathematical model being designed to correspond with experiment only on a macroscopic (and not atomic) level. In this spirit imagine the considered body arbitrarily subdivided into volume elements which are macroscopically small but contain many atoms. The theoretical mass-density is then that function whose integral over any volume element agrees with the actual mass of material there, and whose spatial variation is as smooth analytically as a stipulated large number of different subdivisions will permit. (Of course, in practice, a mass distribution is not determined operationally in this way, and is either simply taken as uniform at the outset or roughly inferred from other observed properties.) With the distribution so defined in principle, the gravitational field of a body may logically be assumed, in the same spirit, as an integral over the whole volume (by contrast with a sum over all atoms)—and, moreover, as an integral of inverse-square contributions. For this is the distance-dependence observed in laboratory experiments with pieces of matter that are small by a macroscopic, and not atomic, standard.

The field associated with any extended body is thus taken as

$$\int_P \frac{(\mathbf{r}_P + \boldsymbol{\rho} - \mathbf{r})}{|\mathbf{r}_P + \boldsymbol{\rho} - \mathbf{r}|^3}\, d\mu, \tag{1}$$

and is asserted to be the independent additional acceleration experienced by a free mass-element in the generic position $\mathbf{r}$ (referred for convenience to the mass-centre G of the system to which the body belongs). The integral extends over the mass distribution of the body, with differential element $d\mu$ at $\boldsymbol{\rho}$ with respect to the mass-centre P at $\mathbf{r}_P$ (Fig. 6). P is defined by the property

$$\int_P \boldsymbol{\rho}\, d\mu = \mathbf{0} \tag{2}$$

analogously to (4) in §1.6; when the body is not rigid P does not usually coincide at all times with a specific material element. Equation (1) generalizes the original gravitational postulate in §1.5, reducing to it when the dimensions of the body are negligibly small.

When **r** is within the material itself, the integrand in (1) becomes infinite at that point, but the improper integral converges and is termed the **self-field.** For the present this concept remains purely formal since the mass-elements of the body are mutually constraining and not free: it acquires dynamical significance only when placed in a wider theoretical framework relating to the internal stress and

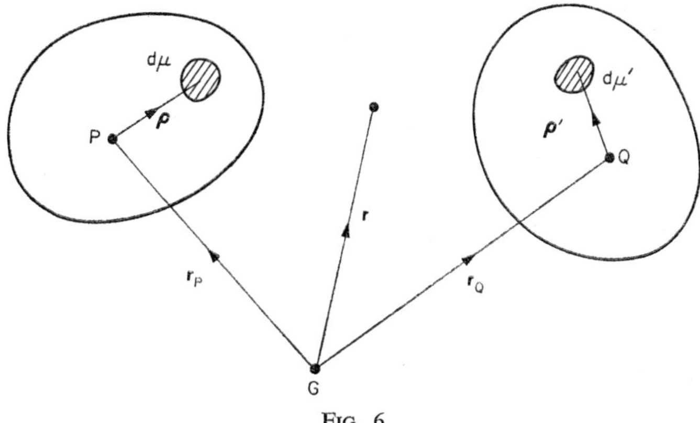

FIG. 6

distortion in a non-rigid body. Here we note merely that, when defined over all space, (1) varies continuously, even across the body surface and any interior surface where there happens to be a finite jump in mass-density.

The movement induced in an extended body itself by a gravitational field, **f** say, due to other matter must now be considered, in particular the acceleration of its mass-centre (rotational and distortional motion of the whole body is treated later in §2.10). For this a further postulate is needed: the mass-centre P of a free body (not necessarily rigid) has an independent additional acceleration $\mathbf{a}_P$ where

$$\mu_P \mathbf{a}_P = \int_P \mathbf{f} d\mu, \tag{3}$$

μ_P being the total mass. This is clearly analogous to, but not inferrable from, the formula for the additional acceleration of the mass-centre of a system of free mass-points. $\mathbf{f}$ can be regarded as involving the self-field or not, at will, since expression (1) integrated with respect to $\mathbf{r}$ over the whole body vanishes identically (this two-fold integral is effectively the sum of mutually-cancelling contributions by all pairs of mass-elements). Note also that $\mathbf{a}_P$ is not generally equal to the local field-value of $\mathbf{f}$ at P.

The contribution to $\mathbf{f}$ by another body with mass-centre Q and total mass μ_Q is given by (1) written with $\boldsymbol{\rho}$ replaced by $\boldsymbol{\rho}'$, $d\mu$ by $d\mu'$, and subscript P by Q (Fig. 6). By (3) the corresponding contribution $\mathbf{a}_{PQ}$ (say) to the acceleration of P is obtained by a further integration over body P:

$$\mu_P \mathbf{a}_{PQ} = \int_P \int_Q \frac{\{(\mathbf{r}_Q - \mathbf{r}_P) + (\boldsymbol{\rho}' - \boldsymbol{\rho})\}}{|(\mathbf{r}_Q - \mathbf{r}_P) + (\boldsymbol{\rho}' - \boldsymbol{\rho})|^3} \, d\mu' d\mu. \tag{4}$$

This contribution is *not* generally directed along PQ. By alternating the roles of the two bodies, the additional acceleration $\mathbf{a}_{QP}$ of Q by body P is obtained from (4) as

$$\mu_Q \mathbf{a}_{QP} = \int_Q \int_P \frac{\{(\mathbf{r}_P - \mathbf{r}_Q) + (\boldsymbol{\rho} - \boldsymbol{\rho}')\}}{|(\mathbf{r}_P - \mathbf{r}_Q) + (\boldsymbol{\rho} - \boldsymbol{\rho}')|^3} \, d\mu d\mu'$$

by interchanging $\boldsymbol{\rho}$ and $\boldsymbol{\rho}'$, $d\mu$ and $d\mu'$, and subscripts P and Q. But this integrand is just the preceding one with the opposite sign and the order of integrations is immaterial. Hence

$$\mu_P \mathbf{a}_{PQ} + \mu_Q \mathbf{a}_{QP} = \mathbf{0}. \tag{5}$$

This is equivalent to the statement that the combined mass-centre of two extended bodies has zero additional acceleration in respect of the mutual (or total) fields.

In fact, this is true for an arbitrary system of free bodies with mass-centres P, Q, ... For the additional accelerations of P, Q, . . . and their combined mass-centre G in the resultant field of the system are given by

$$\left.\begin{aligned} &\mathbf{a}_P = \sum_Q \mathbf{a}_{PQ} \quad \text{with } \mathbf{a}_{PP} = \mathbf{0}, \text{ etc.,} \\ \text{and} \quad &(\Sigma\mu_P)\mathbf{a}_G = \Sigma(\mu_P \mathbf{a}_P) = \sum_{\text{pairs}} (\mu_P \mathbf{a}_{PQ} + \mu_Q \mathbf{a}_{QP}) = \mathbf{0} \end{aligned}\right\} \tag{6}$$

since (5) holds for every pair. The total acceleration of G is therefore

$$\Sigma(\mu_P \mathbf{e}_P)/\Sigma\mu_P \tag{7}$$

in a frame of reference where the total accelerations of the separate mass-centres would be $\mathbf{e}_P + \mathbf{a}_P$, etc.

In the event that the system moves in a uniform external field so that $\mathbf{e}_P = \mathbf{e}_Q = \ldots = \mathbf{e}$, say, expression (7) also reduces to $\mathbf{e}$. That is, G moves as if the whole mass were concentrated there, or as if the separate masses were concentrated at $P, Q, \ldots$. Moreover, the accelerations of $P, Q, \ldots$ relative to G are then just $\mathbf{a}_P, \mathbf{a}_Q, \ldots$ given by (4) with (6) in terms of inverse-square contributions from the system itself. In particular, these remarks hold for the solar system when the frame is Newtonian, generalizing a similar theorem proved in §1.6 under the mass-point idealization.

On the other hand, the collective motions of centres $P, Q, \ldots$ about G do not admit first integrals similar to those for a system of mass-points (§1.10). Thus, the rate of change of 'orbital' moment of momentum about G is

$$\frac{d}{dt}\Sigma(\mathbf{r} \wedge \mu\dot{\mathbf{r}})_P = \Sigma(\mathbf{r} \wedge \mu\ddot{\mathbf{r}})_P = \sum_{\text{pairs}} \{\mu_P(\mathbf{r}_P - \mathbf{r}_Q) \wedge \mathbf{a}_{PQ}\}$$

by (5) and (6) with $\ddot{\mathbf{r}}_P = \mathbf{a}_P$, etc., as just proved when $\mathbf{e}_P = \mathbf{e}_Q$, etc. In general the vector products do not vanish since the mutually induced accelerations in bodies of a pair are not usually in the line of mass-centres. Consequently, when the bodies are not mass-points, the orbital moment of momentum is not generally conserved, by contrast with (4) in §1.10. As we shall see, when a body has extension, its 'rotational' moment of momentum in spinning about its own mass-centre must also be considered; what is conserved is the sum of the orbital and rotational contributions from all bodies of the system (§2.10). A notable (but not unique) exception occurs when all the bodies are spheres, with centro-symmetric density distributions; the mutual accelerations are then always in the line of centres (§1.16) and the orbital and rotational contributions are uncoupled and separately conserved (indeed, the individual spins remain constant).† Similar remarks apply in respect of the integral

† To close approximation the solar system behaves in this manner. Although the two contributions have the same order of magnitude, the astronomical plane defined for the orbital part alone (§1.10) is virtually invariable in a sidereal frame. The rotational moments of momentum of celestial bodies are not known very accurately, but, for example, the value for the sun is about twenty times the *orbital* value for the earth. The 'dynamical invariable plane', for the whole moment of momentum, deviates appreciably from the astronomical plane.

of energy (§2.12). The rate of change of the part of the kinetic energy due to orbital motion about G is

$$\frac{d}{dt}\tfrac{1}{2}\,\Sigma(\mu\dot{\mathbf{r}}^2)_P = \Sigma(\mu\dot{\mathbf{r}}\ddot{\mathbf{r}})_P = \sum_{\text{pairs}}\left\{\mu_P\mathbf{a}_{PQ}\frac{d}{dt}(\mathbf{r}_P - \mathbf{r}_Q)\right\}.$$

But this does not integrate as it stands when the right side is not at all times the rate of change of a function of the configuration of mass-centres [a function such as V in (2) of §1.10 when the bodies are mass-points or uniform spheres for example]. In fact, the accelerations of the mass-centres usually depend as well on the orientations and possible distortions of all bodies.

1.14 Gravitational Potential and Potential Energy

The field of acceleration associated with a single body is given by (1) in §1.13 and can be written as

$$\text{grad}\int_P \frac{d\mu}{|\mathbf{r}_P - \mathbf{r} + \boldsymbol{\rho}|}. \tag{1}$$

At points $\mathbf{r}$ outside the body the equivalence of the two forms is immediate, since both integrands are well-behaved and the gradient with respect to $\mathbf{r}$ can be generated forthwith by differentiating under the integral sign, the region of integration being of course fixed. At points inside the body a formal equivalence is likewise immediate, but the integrands are unbounded; the improper integrals converge but justification of the procedure is not elementary.†

The integral in (1), with sign reversed, is known as the **gravitational potential.** It is a continuous function of the generic position $\mathbf{r}$, and implicitly of time through its dependence on the varying configuration. Any gravitational field whatever is naturally attributable to some system of bodies and so derivable as the vector gradient of a potential function of $\mathbf{r}$:

$$\mathbf{f} = -\,\text{grad}\,\phi \quad \text{where } \phi(\mathbf{r}) = -\sum\int_P \frac{d\mu}{|\mathbf{r}_P - \mathbf{r} + \boldsymbol{\rho}|}. \tag{2}$$

† It is given in broad outline in specialist texts on gravitational theory, for example A. S. Ramsey: *Introduction to the Theory of Newtonian Attraction* (Cambridge University Press 1952). For a full treatment O. D. Kellogg: *Foundations of Potential Theory* (Springer 1929) should be consulted.

This is just the sum of the potentials of the individual fields (including the self-field of any body containing the point $\mathbf{r}$). From (3) of §1.13 the equation of motion of a typical mass-centre is given by

$$\mu_P \mathbf{a}_P = -\int_P \operatorname{grad} \phi d\mu, \tag{3}$$

alternatively to (6) in §1.13.

The total potential ϕ is to be sharply distinguished from the total potential energy V (the clash in terminology being unfortunate, but firmly established). ϕ is a function of the independent variable $\mathbf{r}$ as well as the entire mass-configuration, while V has to do only with the latter and is defined as an overall integral:

$$V = -\sum_{\text{pairs}} \int_P \int_Q \frac{d\mu\, d\mu'}{|(\mathbf{r}_P - \mathbf{r}_Q) + (\boldsymbol{\rho} - \boldsymbol{\rho}')|} = \tfrac{1}{2}\Sigma \int_Q \phi d\mu' \tag{4}$$

where ϕ in a typical body Q is given by (2) with $\mathbf{r} = \mathbf{r}_Q + \boldsymbol{\rho}'$. The factor $\frac{1}{2}$ is needed in the second expression for V since a typical inverse distance (say between $d\mu$ and $d\mu'$ in bodies P and Q) appears twice: once in the integral over P of the part of ϕ due to Q and again in the integral over Q of the part of ϕ due to P.

For future reference we note, in passing, a variation formula:

$$\Sigma \int_P \mathbf{f}\delta(\mathbf{r} + \boldsymbol{\rho})d\mu + \delta V = 0 \tag{5}$$

for arbitrary infinitesimal changes in the geometry (including shapes and orientations). This may be proved most conveniently by computing the variation of the integrand in the first expression for V, coupled with the observation that the range of integration over the *mass* distribution obviously does not alter even though the positions and shapes of material elements may do. With the special choice of variations such that the bodies are merely translated independently, without rotation or distortion, $\delta(\mathbf{r} + \boldsymbol{\rho}) = \delta\mathbf{r}_P$ throughout body P for example and (5) becomes

$$\Sigma(\mu \mathbf{a} \delta \mathbf{r})_P + \delta V = 0 \tag{6}$$

with the help of (3) in §1.13. This generalizes a similar result, (9) in §1.6, for mass-points.

When a body is rigid, the contribution to V from its self-energy is invariable, and is by convention omitted since it has no dynamical significance. In particular, for a system of rigid bodies idealized as mass-points (for which, as it happens, the constant self-energies are unbounded in the limit), definition (4) is replaced by

$$\left.\begin{aligned} V &= -\sum_{\text{pairs}} \{\mu_Q \mu_R / |\mathbf{r}_Q - \mathbf{r}_R|\} = \tfrac{1}{2}\Sigma \mu_P \phi_P \\ \text{where} \qquad \phi_P &= -\sum_{Q \neq P} \{\mu_Q / |\mathbf{r}_P - \mathbf{r}_Q|\} \end{aligned}\right\} \qquad (7)$$

is the potential at P due to the other bodies. In place of (3) one then has

$$\mathbf{a}_P = -\operatorname{grad}_P \phi \qquad (8)$$

for the additional acceleration of P, where this notation signifies the formal gradient of ϕ_P in (7) when regarded as a function only of the coordinates of P. The connexion with (8) in §1.6 is established by verifying that $\operatorname{grad}_P V = \mu_P \operatorname{grad}_P \phi$.

1.15 Some Properties of Gravitational Fields

We prove here a few general theorems that will be needed subsequently in calculating fields of particular bodies.

The first theorem is a simple consequence of the inverse-square dependence, and relates to the **gravitational flux**. By definition the flux of any field $\mathbf{f}$ through an infinitesimal area dS with unit normal $\mathbf{n}$ is $\mathbf{fn}dS$; this may be regarded either as the product of dS with the normal component of $\mathbf{f}$, or as the product of the magnitude of $\mathbf{f}$ with the projection of dS at right-angles to $\mathbf{f}$. To begin with, consider just the centro-symmetric field of a mass-point P and imagine an infinitesimal cone (not necessarily circular) drawn from P. Take for the infinitesimal area an arbitrarily inclined section of the cone at distance r from P. Since the field is radial and proportional to $1/r^2$ while the right section of the cone is proportional to r^2, the flux has the same magnitude for all such sections at any point of the cone. Next consider the net flux through any closed regular surface S in the direction of its outward normal. A typical cone with vertex at P, extended indefinitely in both directions, cuts S in an even number of elementary areas; their contributions to the flux have equal magnitude (as just proved) but are positive or negative according as the field direction enters or leaves S. The net flux through S therefore has one of two fixed values: if S does not contain P the flux is zero; if S does contain P the flux is non-zero and its universal value can be identified from a particular S. Through a sphere centred on P the inward flux is obviously $4\pi d\mu$, where $d\mu$ is the mass, since the surface area is $4\pi r^2$ and the field is $d\mu/r^2$ radially inward.

Finally, for any extended distribution of matter (which may be part or all of a single body or of a system),

$$\int \mathbf{fn}dS = -4\pi \int_S d\mu \tag{1}$$

where $\mathbf{f}$ is the resultant field, $\mathbf{n}$ is the unit outward normal, and the mass integral extends over the matter inside S. Since the field is continuous, the result holds even when S passes through matter; this is easily established by introducing auxiliary surfaces on either side of S and tending to it, and then applying (1) to the field of matter within the inner surface and separately to the field of matter outside the outer surface. A statement of (1) in words is: any field of additional acceleration has a total inward flux, through any closed surface, equal to $4\pi \times$ the mass of that matter which both contributes to the field and is enclosed by the surface.

Now, by the divergence theorem, (1) may be written as an integral over the volume v within S:

$$\int(\operatorname{div}\mathbf{f} + 4\pi\sigma)dv = 0$$

where σ is the mass-density such that $\sigma dv = d\mu$. The transformation is permissible since $\mathbf{f}$ is continuous, and its spatial derivatives may be shown piecewise continuous when σ is analytic (the jumps occurring where σ is discontinuous). By applying this result to a vanishingly small region about an arbitrary point (but not on a jump surface) we deduce that

$$\operatorname{div}\mathbf{f} + 4\pi\sigma = 0 \tag{2}$$

where σ is, specifically, the local density of matter contributing to the field $\mathbf{f}$. Conversely, the flux theorem (1) holds for any field with the property (2). With $\mathbf{f} = -\operatorname{grad}\phi$, (2) shows that the potential satisfies a second-order partial differential equation of Poisson type wherever σ is an analytic function of position:

$$\nabla^2\phi = 4\pi\sigma \tag{3}$$

where ∇^2 is the Laplacian operator. This offers another way of calculating the potential, instead of integrating directly over a distribution of mass as in (2) of §1.14. This alternative is especially convenient when the distribution is unspecified or indeed unknown, as happens in celestial mechanics, but in compensation the variation of the potential is given, or can be inferred, over surfaces containing undefined matter.

The method is made precise by a uniqueness theorem. Let S denote, collectively, the closed surfaces where the potential is given (presumed consistently with *some* mass distribution so that the problem is assured of a solution). The actual potential has the following properties (among others): (i) ϕ takes the assigned boundary-values; (ii) ϕ is of order $1/r$ at distances large compared with the system dimensions; (iii) ϕ and its gradient are continuous, while the second derivatives are at least piecewise continuous; (iv) equation (3) is satisfied everywhere outside S. We wish to show that these properties are possessed by no more than one function (which can only be the required potential). Let the prefix Δ signify the difference of corresponding quantities in a supposedly distinct pair of solutions. In view of (iii) and by Green's theorem†

$$\int(\Delta\phi)(\mathbf{n}\Delta\mathbf{f})d(S+\Sigma) = \int(\Delta\phi \operatorname{div} \Delta\mathbf{f} + \Delta\mathbf{f} \operatorname{grad} \Delta\phi)dv \tag{4}$$

where v is the volume between S and any distant closed surface Σ, and $\mathbf{n}$ is the outward unit normal from v. Now $\Delta\phi = 0$ on S by (i), $\Delta\phi$ is of order $1/r$ and $\Delta\mathbf{f}$ of order $1/r^2$ by (ii), and $\operatorname{div} \Delta\mathbf{f} = 0$ in v by (iv). Since the area of Σ is of order r^2, the corresponding integral is of order $1/r$ and so tends to zero when Σ recedes indefinitely. In the limit only the last term in (4) survives and yields

$$\int(\Delta\mathbf{f})^2 dv = 0$$

† It is recalled that this is obtained by integrating the elementary identity

$$\operatorname{div}(\alpha\mathbf{a}) = \alpha \operatorname{div} \mathbf{a} + \mathbf{a} \operatorname{grad} \alpha$$

over a volume, and transforming the left-hand side to a surface integral by the divergence theorem. Equation (4) has $\alpha = \Delta\phi$, $\mathbf{a} = \Delta\mathbf{f}$.

where the integration extends over the whole of space outside S. Since the integrand here is continuous and non-negative it follows that $\Delta\mathbf{f} = \mathbf{0}$ and thence that $\Delta\phi = 0$ since it vanishes on S.

In retrospect it is apparent that uniqueness can be proved for any boundary conditions such that the integral over S in (4) vanishes. This clearly happens, for instance, when *either* the potential *or* the normal component of the field is prescribed over different portions of S. It happens also when the variation of potential over each disjunct part of S is given only to within an unspecified additive constant, but the total mass inside every such part is known. For the integral reduces to a linear combination of the respective fluxes of $\Delta\mathbf{f}$, and these vanish separately since the fluxes of $\mathbf{f}$ are fixed by the total masses.

1.16 Mutual Fields of Distant Bodies

We are now in a position to resume the analysis of gravitational fields of extended bodies. To begin with, consider a sphere of radius r_0 in which the mass-density is any piecewise continuous function $\sigma(r)$ of the radial distance r (centro-symmetric distribution). By symmetry the field is radial and so the flux theorem of §1.15 suffices by itself to determine the field magnitude:

$$\left.\begin{aligned} f(r) &= \frac{\mu}{r^2}, \quad r \geqslant r_0; \\ f(r) &= \frac{1}{r^2}\int_0^r d\mu \quad \text{with } 4\pi r^2\sigma = \frac{d\mu}{dr}, \quad r \leqslant r_0. \end{aligned}\right\} \tag{1}$$

The internal field varies from zero at the centre to μ/r_0^2 at the surface, though not necessarily monotonically; if σ is uniform, $f = \mu r/r_0^3$ and $\phi = \mu(r^2 - 3r_0^2)/2r_0^3$. Irrespective of the $\sigma(r)$ variation, however, the external field is everywhere equivalent to that of a mass-point situated at the centre of the sphere (*Principia:* Book I, Prop. LXXIV, by synthetic geometry). The present proof appears to be the simplest possible; by contrast, the use of Poisson's equation, for example, would be too sophisticated here as well as less illuminating. The result is remarkable in that nothing similar holds for inverse powers other than the second.

If the earth could be treated as centro-symmetric its gravitational field at its mean radius r_E would be

$$r_E \left(\frac{2\pi}{\tau}\right)^2 \left(\frac{r_M}{r_E}\right)^3 \bigg/ \left(1 + \frac{\mu_M}{\mu_E}\right)$$

in terms of the period τ and mean radius r_M of the moon's orbit and the mass-ratio μ_M/μ_E [cf. (6) in §1.7]. With $\tau = 27{\cdot}3217 \times 86{,}400$ seconds, $r_M/r_E = 60{\cdot}337$, $\mu_M/\mu_E = 0{\cdot}0123$, $r_E = 6371$ km, this formula gives $979{\cdot}3$ cm/sec^2, which is satisfactory having regard to the simplifying assumptions (see §1.18). Historically, this calculation was crucial as a test of the universal validity of the hypothesis

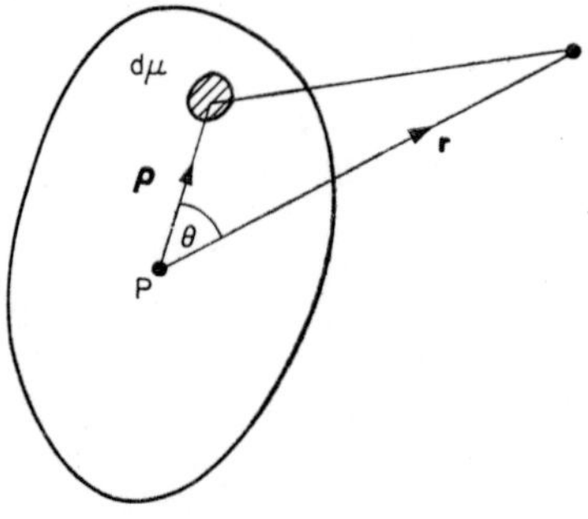

FIG. 7

that all bodies, and not only widely separated ones, have mutual motions in accord with the inverse-square law. Having conjectured the field of a solid sphere, Newton first tried the test in 1666, and again in 1672 with an improved value for the earth's radius. [*Principia:* Book III, Prop. IV; *The System of the World*, §§10 and 11. The then unknown mass-ratio was over-estimated by a factor of nearly 2 on the supposition that the earth and moon had equal densities.] It is believed that Newton's announcement of the principle of universal gravitation was nevertheless deferred until the conjecture itself had been rigorously proved.

The additional acceleration of the centre P of one centro-symmetric body by another with centre Q follows from (3) of §1.13. **f** there is obtained by concentrating μ_Q at Q and its integral over body P is then $-\mu_Q \times$ the field of body P at Q, or $\mu_P\mu_Q/|PQ|^2$ in the direction from P to Q. Whence $\mathbf{a}_P$ is $\mu_Q/|PQ|^2$ towards Q

and similarly $\mathbf{a}_Q$ is $\mu_P/|PQ|^2$ towards P (*Principia:* Book I, Prop. LXXVI). And, generally, the translational motions of any system of centro-symmetric spheres are precisely as if their masses were concentrated at their respective centres. The equations of motion therefore have first integrals of the type holding for mass-points, as foreshadowed in §1.13.

Consider, next, a body (or system) with arbitrary shape and density distribution. We begin by investigating the distant field, since this is of greatest interest in celestial mechanics. Let $\mathbf{r} = r\mathbf{l}$ be measured from the mass-centre P to a generic external position. A typical mass element $d\mu$ in the body is specified by spherical polar coordinates (ρ, θ, ψ) with respect to P and relative to an axis in direction $\mathbf{l}$ (Fig. 7). The potential function is then expressible as

$$\phi(\mathbf{r}) = -\int_P \left(1 - \frac{2\rho}{r}\cos\theta + \frac{\rho^2}{r^2}\right)^{-\frac{1}{2}} \frac{d\mu}{r}$$

integrated over the body, by the elementary formula for the side of the triangle opposite θ. The radical is now expanded as a power series in the ratio ρ/r:

$$\phi(\mathbf{r}) = -\int_P \left[1 + \frac{\rho}{r} P_1(\theta) + \frac{\rho^2}{r^2} P_2(\theta) + \ldots\right] \frac{d\mu}{r} \tag{2}$$

where the coefficients $P_1, P_2, \ldots$ are (by inspection) polynomials in $\cos\theta$ of degrees indicated by their subscripts. For instance, the first two coefficients are readily found to be

$$P_1 = \cos\theta, \qquad P_2 = \tfrac{1}{2}(3\cos^2\theta - 1). \tag{3}$$

In fact, these are the wellknown Legendre polynomials such that all $\rho^n P_n(\theta)$ are solutions of Laplace's equation in the variable $\boldsymbol{\rho}$ with coordinates (ρ, θ, ψ), while $P_n(\theta)/r^{n+1}$ are solutions in the variable $\mathbf{r}$ with coordinates r, θ, and an azimuthal angle around any fixed $\boldsymbol{\rho}$ direction. The bracketed series in (2) is uniformly convergent for all real θ provided $\rho/r < 1$; this is established by regarding it as the product of the binomial expansions of $(1 - xe^{i\theta})^{-\frac{1}{2}}$ and $(1 - xe^{-i\theta})^{-\frac{1}{2}}$ in $x = \rho/r$, which both converge absolutely when $|x| < 1$. The integrated series therefore converges at least outside a spherical surface, centre P, which just encloses the body (and even converges right up to the body, by the principle of analytic continuation). Also, by the previously mentioned property of Legendre polynomials, every term in the resulting expansion for ϕ is a solution of Laplace's equation in the variable $\mathbf{r}$. In passing, it may be observed that the preceding result for the external field of a centro-symmetric sphere is recovered with the help of another property: the surface integral of any $P_n(\theta)$ vanishes over all concentric spheres.

With P defined as the mass-centre, $\int \rho \cos \theta d\mu$ is the projection of $\int \boldsymbol{\rho} d\mu$ on the direction $\mathbf{l}$ and vanishes by (2) of §1.13. The first terms in the expansion are thus

$$\phi(\mathbf{r}) = -\left[\frac{\mu}{r} + \frac{1}{r^3}\int \rho^2 P_2(\theta) d\mu + \ldots\right].$$

The leading term would be the potential if the body (or system) were treated as a mass-point at P. The second term can be put in an explicit form due to MacCullagh (1847). For this purpose we re-write P_2 in (3) as $\frac{1}{2}(2 - 3 \sin^2 \theta)$ and define

$$I(\mathbf{l}) = \int (\rho \sin \theta)^2 d\mu, \qquad J = \int \rho^2 d\mu. \tag{4}$$

J is a constant, the polar moment of inertia about P, analogous to the quantity already introduced for a system of mass-points in §1.11. $I(\mathbf{l})$ is the (axial) moment of inertia about the direction $\mathbf{l}$ and is a function of the components of $\mathbf{l}$ on axes fixed in the body. MacCullagh's formula is therewith

$$\phi(\mathbf{r}) = -\left[\frac{\mu}{r} + \frac{2J - 3I(\mathbf{l})}{2r^3} + O\left(\frac{1}{r^4}\right)\right], \quad \mathbf{r} = r\mathbf{l}. \tag{5}$$

For the first-order correction to the mass-point idealization it is thus necessary to know only the overall inertias with respect to P, and not the detailed distribution of density; conversely, the inertias can be deduced for any celestial body from an analysis of perturbations of the elliptical orbits of its satellites or neighbouring planets. For the earth–moon system some later terms in the expansion also have measurable effects, in regard both to their mutual motions and the orbits of space vehicles.

The distant field of acceleration is the negative gradient of expression (5):

$$\left.\begin{aligned} \mathbf{f} &= -\frac{\mu}{r^2}(\mathbf{l} + \boldsymbol{\alpha}) + O\left(\frac{1}{r^5}\right) \\ \text{where} \quad \boldsymbol{\alpha} &= \frac{3(2J - 3I)}{2\mu r^2}\mathbf{l} + \frac{3}{2\mu r}\operatorname{grad} I(\mathbf{l}). \end{aligned}\right\} \tag{6}$$

The first-order correction to the inverse-square law thus consists of a component along $\mathbf{l}$ together with a transverse component in the direction of grad $I(\mathbf{l})$ (this gradient is of order $I(\mathbf{l})/r$ at most and has zero radial component since $I(\mathbf{l})$ is independent of r). For convenience we defer detailed calculations until the character of the dependence has been explored (§3.4), and merely note here the existence of the transverse acceleration (cf. the discussion at the end of §1.13). Next, consider a second body situated in the field, with mass μ' and inertias I', J'. The additional acceleration of its mass-centre is obtained by integration of (6) over this body, by (3) of §1.13. The integral of the inverse-square term in (6) is $-\mu\mathbf{f}'$, where $\mathbf{f}'$ is the field of the second body at the mass-centre of the first:

$$\mathbf{f}' = \frac{\mu'}{r^2}(\mathbf{l} + \boldsymbol{\alpha}') + O\left(\frac{1}{r^5}\right),$$

where
$$\boldsymbol{\alpha}' = -\frac{3(2J' - 3I')}{2\mu' r^2}\mathbf{l} + \frac{3}{2\mu' r}\,\text{grad}\, I'(\mathbf{l})$$

and $\mathbf{r}$ is now the vector distance between centres; the integral of the correction term is just $-\mu\mu'\boldsymbol{\alpha}/r^2$ to the required order of approximation. In all,

$$\mu\mathbf{a} = -\mu'\mathbf{a}' = \frac{\mu\mu'}{r^2}(\mathbf{l} + \boldsymbol{\alpha} + \boldsymbol{\alpha}') + O\left(\frac{1}{r^5}\right), \tag{7}$$

by recalling (5) of §1.13 or by alternating roles.

The characteristic property of centro-symmetric bodies can now be seen in perspective. At sufficiently large distances, and with correspondingly small error, this property is shared by any mass-distribution such that the moment of inertia $I(\mathbf{l})$ is independent of direction. For then $I(\mathbf{l}) = \frac{2}{3}J$, since $2J$ is the sum of the moments of inertia about any orthogonal triad of directions, and the leading correction $\boldsymbol{\alpha}$ vanishes. Such a distribution of matter is said to possess all-round kinetic symmetry with respect to its mass-centre. Actually, there is a special sub-class of these bodies whose field at all external points is *exactly* as if the mass were concentrated at the mass-centre ('centrobaric body').†

† The topic is somewhat academic; details are given by W. D. MacMillan: *The Theory of the Potential* (Dover Publications 1958).

1.17 Fields of Axisymmetric Bodies

In the preceding analysis the geometry and density variation are unrestricted but the final formula of MacCullagh tends to become less exact in the neighbourhood of the body. In a complementary approach, now to be described, the approximation is uniform everywhere but the density and geometry are required to be nearly centro-symmetric. Also, instead of direct integration over the mass distribution, the potential is generated by solving Laplace's equation with boundary values over the body surface.

Many celestial bodies (the moon being a notable exception) may be supposed to have an axis of symmetry. Take spherical polar coordinates (r, θ, ψ) with respect to this axis and the mass-centre. The equation of the surface S can be represented to any desired accuracy by a linear combination of Legendre polynomials:

$$r(\theta) = r_0[1 - \Sigma \varepsilon_\nu P_\nu(\theta)] \quad \text{on} \quad S \tag{1}$$

where (by the initial requirement) the coefficients ε_ν are all small compared with unity, r_0 is the mean radius averaged over S, and the summation extends over a sufficient number of positive integers ν. The external potential associated with this body is an axially-symmetric solution of Laplace's equation differing slightly from a simple inverse-distance relation, and is therefore representable in the form

$$\phi(r, \theta) = -\frac{\mu}{r}\left[1 - \Sigma \eta_\nu \left(\frac{r_0}{r}\right)^\nu P_\nu(\theta)\right] \tag{2}$$

where μ and the small coefficients η_ν remain to be determined. The radial and transverse components of the field of acceleration are then obtained from

$$\left.\begin{aligned} \frac{\partial \phi}{\partial r} &= \frac{\mu}{r^2}\left[1 - \Sigma(\nu + 1)\eta_\nu \left(\frac{r_0}{r}\right)^\nu P_\nu(\theta)\right], \\ \frac{1}{r}\frac{\partial \phi}{\partial \theta} &= -\frac{\mu}{r^2}\Sigma \eta_\nu \left(\frac{r_0}{r}\right)^\nu \frac{d}{d\theta} P_\nu(\theta). \end{aligned}\right\} \tag{3}$$

According to the theorem in §1.15 the flux through every surface containing the body is the same; its value here is precisely $4\pi\mu$, as can be seen immediately by integrating (3) over any sphere and recalling that such integrals of Legendre polynomials vanish. Consequently, as the notation anticipated, μ must be the total mass of the body. If desired, the other coefficients in (2) can be identified as integrals over the density distribution, by comparison with the equivalent expansion in inverse powers of r in §1.16 (note the different uses of θ). In particular, the coefficient of $1/r^3$ is given explicitly by MacCullagh's formula, (5) in §1.16:

$$\eta_2 \mu r_0^2 P_2(\theta) = \tfrac{3}{2} I(\theta) - J.$$

That the right side is indeed proportional to the second Legendre polynomial, (3) in §1.16, is verifiable from the relations

$$2J = C + 2A, \quad I(\theta) = C\cos^2\theta + A\sin^2\theta,$$

which are proved later (§3.4). Here $C \equiv I(0)$ is the moment of inertia about the axis of symmetry and $A \equiv I(\tfrac{1}{2}\pi)$ is the moment of inertia about any transverse direction through the mass-centre (in gravitational units). Whence

$$\eta_2 = (C - A)/\mu r_0^2. \tag{4}$$

Our present concern, however, is with the potential and its normal gradient on S itself. These are

$$\phi(\theta) = -\frac{\mu}{r_0}[1 + \Sigma(\varepsilon_\nu - \eta_\nu)P_\nu(\theta) + \ldots], \tag{5}$$

$$\frac{\partial\phi}{\partial n} = \frac{\mu}{r_0^2}[1 + \Sigma\{2\varepsilon_\nu - (\nu + 1)\eta_\nu\}P_\nu(\theta) + \ldots], \tag{6}$$

where the unwritten terms are of the order of the squares and products of the small coefficients, and the derivative $\partial/\partial n$ along the outward normal can correspondingly be approximated by $\partial/\partial r$. If, now, the equation of S and the variation of potential over it were given by (1) and (5), with any particular values of the coefficients, the external potential could be written down at once from (2), without direct knowledge of the density distribution within S. This procedure is justified by the uniqueness theorem in §1.15 since (2) is a harmonic function satisfying the requisite boundary conditions on S and at infinity. Similar remarks apply when the equation of S with the local normal acceleration (5) are given.

For actual celestial bodies the situation is not so straightforward.

Further progress is made possible by an added hypothesis: the present surface is broadly the same (superficial features apart) as it was at an epoch when the body might be presumed to have been molten and to have had the same spin as now. More precisely, the present shape is the steady-state configuration of a spinning mass of non-uniform inviscid fluid under its gravitational self-field (oscillating bodily tides due to fields of other bodies being treated separately as a superimposed second-order effect: this is well justified except for close binary stars). With any assumed density variation the actual shape for a given spin can be calculated on this basis, and leads to 'theories of the internal field' instituted by Maclaurin (1740), for uniform density, and Clairaut (1743) in full generality.†

For present purposes, however, it is enough to draw from the hypothesis the immediate conclusion that

$$\phi - \tfrac{1}{2}\Omega^2 r^2 \sin^2\theta = \text{constant on } S, \tag{7}$$

irrespective of the internal density variation, where Ω is the spin about the axis of symmetry relative to a sidereal frame. This formula depends only on the remark that a surface mass-element, when ideally fluid, has exactly the same *tangential* component of acceleration in the meridian plane as a *free* mass-element would have if placed in that position; in other words, the presence of neighbouring fluid elements affects only the normal component. That is, the tangential derivative $\partial\phi/\partial s$ is to be identified with $\Omega^2 r \sin\theta\,\partial(r\sin\theta)/\partial s$ since the actual steady-state acceleration is $\Omega^2 r\sin\theta$ centripetally towards the axis of spin; equation (7) follows by integration along the surface profile.

If, now, the shape (1) is pre-supposed, the local potential is specified by (7) as a function of θ to within an additive constant. This constant, together with the unknown coefficients η_ν, are identified on comparison with (5), the total mass being given. Equation (3) then supplies the external field. Once again, the validity of the method is safeguarded by the uniqueness theorem of §1.15, with the final form of boundary conditions. The analysis is next carried through in detail for the earth.

† Accounts can be found in H. Jeffreys: *The Earth* (Cambridge University Press 1959, 4th edition); W. A. Heiskanen and F. A. Vening Meinesz: *The Earth and its Gravity Field* (McGraw-Hill 1958).

1.18 Gravity and the Figure of the Earth

Except when extreme accuracy is called for, two simplifications are feasible. To begin with, the equatorial plane through the mass-centre may be supposed a plane of symmetry, so that the expansion involves no Legendre polynomials of odd order since these are odd functions of $\cos\theta$. More particularly, it is known that the earth's figure may be closely represented by an oblate spheroid of small eccentricity, apart from fortuitous and superficial topographical features; a spheroidal surface is actually predicted for a slowly-spinning mass of gravitating fluid by the internal field theory mentioned previously. A meridian section with equatorial radius a and polar radius c is the ellipse

$$\frac{1}{r^2} = \frac{\cos^2\theta}{c^2} + \frac{\sin^2\theta}{a^2}.$$

When expanded in powers of the ellipticity $\varepsilon = (a - c)/a$,

$$r(\theta) = r_0[1 - \tfrac{2}{3}\varepsilon P_2(\theta) + O(\varepsilon^2)] \text{ with } r_0 = a(1 - \tfrac{1}{3}\varepsilon), \qquad (1)$$

by (3) in §1.16. Evidently these are also the leading terms in the Legendre expansion, to the same order of approximation, with r_0 as mean radius.

Secondly, the earth's spin is so small that the dominant terms in the Legendre expansion of ϕ over the surface, as obtained from (7) in §1.17 on the fluid model, are

$$\phi(\theta) = -\frac{\mu}{r_0}[1 + \tfrac{2}{3}\beta P_2(\theta) + O(\varepsilon^2)] \text{ where } \beta = \tfrac{1}{2}r_0^2\Omega^2 \Big/ \frac{\mu}{r_0}. \qquad (2)$$

The introduced parameter β, which will later be found to be less than ε, can be regarded as an order-of-magnitude measure of the ratio of centripetal to gravitational acceleration at the equator, or of the greatest proportionate contribution to the potential by the spin term. The ensuing analysis, in which products and powers of ε and β are consistently omitted, is known as the 'first-order theory of the external field' (Clairaut 1743).

By comparing (1) and (2) with the corresponding expansions in §1.17 we find $\varepsilon_2 \simeq \frac{2}{3}\varepsilon$, $\eta_2 \simeq \frac{2}{3}(\varepsilon - \beta)$, and can re-write (2) and (4) there as

$$\left.\begin{aligned} \phi(r,\theta) &= -\frac{\mu}{r}\left[1 + (\varepsilon - \beta)\left(\frac{r_0}{r}\right)^2 (\tfrac{1}{3} - \cos^2\theta) + \ldots\right] \\ \text{with } \varepsilon - \beta &= 3(C - A)/2\mu r_0^2. \end{aligned}\right\} \qquad (3)$$

The function defined on the left of (7) in §1.17, which for the earth is called the **geopotential** (Φ say) is thus

$$\Phi(r, \theta) = -\frac{\mu}{r}\left[1 + (\varepsilon - \beta)\left(\frac{r_0}{r}\right)^2(\tfrac{1}{3} - \cos^2\theta) + \beta\left(\frac{r}{r_0}\right)^3(1 - \cos^2\theta) + \dots\right] \quad (4)$$

when Ω is eliminated in favour of β.

Now, what is commonly known as the 'acceleration due to gravity', always denoted by $\mathbf{g}$, is basically defined as the negative gradient (with respect to the above coordinates) of the geopotential at a considered station, whether on the earth or in space:

$$\mathbf{g} = -\operatorname{grad}\Phi \quad \text{where } \Phi = \phi - \tfrac{1}{2}\Omega^2 r^2 \sin^2\theta. \quad (5)$$

The difference between gravity and the earth's field $\mathbf{f} = -\operatorname{grad}\phi$ is thus

$$\mathbf{f} - \mathbf{g} \equiv \Omega^2 r \sin\theta \text{ centripetally.} \quad (6)$$

Ordinarily, of course, $\mathbf{g}$ is thought of as the acceleration of a mass-point *in vacuo* relative to terrestrial axes of reference; actually, as we shall see later (§2.2), this definition needs to be made unambiguous by stipulating that the observation is made when the mass-point momentarily appears at rest with respect to these axes. Accepting this, and recalling that $\mathbf{f}$ is the acceleration relative to a sidereal frame and with respect to the earth's centre, the equivalence of (6) with the ordinary definition is clear.

It will also be shown later that the direction of a plumb-line at rest in the terrestrial frame coincides with that of local gravity (§2.3). By (5) this is normal to the local Φ level-surface; in particular, for the level surface S in the fluid model, its inclination to the equatorial plane defines the **geographic latitude** λ (Fig. 8). This is to be distinguished from **geocentric latitude** $\frac{1}{2}\pi - \theta$, the connexion being

$$\lambda - (\tfrac{1}{2}\pi - \theta) = \left(\frac{1}{r}\frac{\partial r}{\partial\theta}\right)_S = \varepsilon \sin 2\theta + O(\varepsilon^2) \quad (7)$$

from (1) by simple geometry. The maximum deviation is about 11′ in latitude 45°. A short calculation shows that the **gravitational**

latitude, defined by the direction of the field **f**, is $\lambda - \beta \sin 2\theta$ very nearly, which differs from λ by about 5′ at most (Fig. 8).

The magnitude of **g** is effectively $\partial\Phi/\partial r$, which on S itself is

$$g = \frac{\mu}{a^2}[(1 + \varepsilon - 3\beta) + (5\beta - \varepsilon)\sin^2\lambda] + O(\varepsilon^2),$$

while $f = g(1 + 2\beta\cos^2\lambda)$ to the same order. Introducing the equatorial and polar values

$$g_e = \frac{\mu}{a^2}(1 + \varepsilon - 3\beta), \quad g_p = \frac{\mu}{a^2}(1 + 2\beta), \tag{8}$$

we can write

$$g = g_e\cos^2\lambda + g_p\sin^2\lambda + O(\varepsilon^2). \tag{9}$$

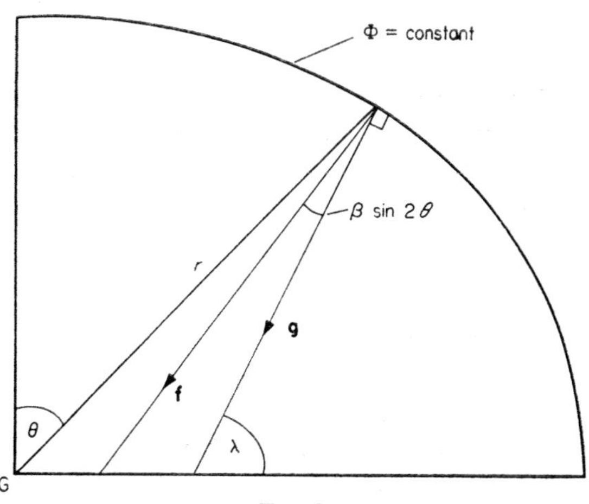

FIG. 8

By fitting this relation to gravimetric data, corrected for station error† and height above mean sea-level, it is found that

$$g_p = 983{\cdot}217 \text{ cm/sec}^2, \quad g_e = 978{\cdot}039 \text{ cm/sec}^2,$$
$$(g_p - g_e)/g_e \simeq 5\beta - \varepsilon \simeq 0{\cdot}005294.$$

Now, from (2) and (8) with consistent approximation,

$$\beta \simeq \tfrac{1}{2}a\Omega^2/g_e = 0{\cdot}0017339$$

† The actual plumb-line direction defines **astronomical latitude** and is affected by local surface features: the deviation from the normal to the reference spheroid is at most 1′ or so.

with $\Omega = 2\pi$ radians/sidereal day (§1.3) $= 7{\cdot}292115\ldots \times 10^{-5}$ radians/sec and $a = 6378{\cdot}388$ km (see below). Whence $\varepsilon = 0{\cdot}003376$, the last decimal place being only nominally significant. In passing, it may be noted that the direct centripetal effect of the earth's rotation is responsible for a part 2β of the total fractional variation of gravity from equator to pole, or very nearly $\frac{2}{3}$ by proportion.†

A check on the internal consistency of the theory is provided by an independent estimate of the difference $(\varepsilon - \beta)$, and so of $(C - A)$ also, obtained from the observed perturbations of the moon's orbit arising from the $P_2(\theta)$ dependence in the earth's field (3):

$$\mathbf{f} = \frac{\mu}{r^2}\left[(-1, 0) + (\varepsilon - \beta)\left(\frac{r_0}{r}\right)^2 (3\cos^2\theta - 1, \sin 2\theta)\right] + \ldots \quad (10)$$

By analogy this formula is valid for any celestial body with axial symmetry. The components, separated by the comma, are respectively along the outward radius and transversely in a meridian plane in the direction of increasing θ. Evidently the earth is not centrobaric since the transverse component vanishes only on the axis of symmetry and in the equatorial plane. Orbits of artificial satellites provide still more accurate estimates of $(\varepsilon - \beta)$, since $(r_0/r)^2$ is relatively much greater.†† By contrast, at planetary distances, the deviation in *magnitude* from the inverse-square dependence is utterly negligible: for example, since the earth's radius is about 4×10^{-5} a.u., the fractional deviation at the solar distance 1 a.u. is only of order 10^{-11}. On the other hand, the deviation in *direction*, small though it is, is associated with observable rotational effects, among them the equinoctial precession [§3.8 (v)].

Values of the ellipticity obtained by all methods are within 0·5 per cent of a nominal value adopted by international agreement in 1924:

$$\varepsilon = 1/297 \text{ (exactly)} = 0{\cdot}003367\ldots$$

† So-called 'standard gravity' has been defined (1950) as 980·665 cm/sec² (= 32·1740 ft/sec²). This is not far from the value 980·628 cm/sec² in latitude 45° at sea-level, given by (9).

†† The presence of a P_3 term is also indicated, implying a very slight asymmetry about the equatorial plane; coefficients of a few subsequent terms can be estimated as well. The usual orbit can be broadly described as an ellipse that slowly turns in its own plane while this precesses around the earth's axis; these rotations are both of order $(\varepsilon - \beta)(r_0/l)^2$ radians per revolution of the satellite, where l is the semi-latus rectum of the orbit.

together with

$$a = 6378{\cdot}388 \text{ km}, \qquad c = 6356{\cdot}912 \text{ km}.$$

Geodetic surveys have established that this standard spheroid of reference is coincident, to within a few metres, with mean sea-level in the oceans. For comparison it may be mentioned that a gravitating mass of *uniform* fluid with small spin Ω would have an ellipticity of $5\beta/2$ or about $1/230$ (*Principia:* Book III, Prop. XIX). Since this exceeds the above value, the earth must be appreciably more dense towards its centre, in agreement with inferences from seismology (cf. §2.6).

Finally, the gravitational mass of the earth, in metric units, can be deduced from the first of (8) with the data for the reference spheroid and the measured gravity at the equator:

$$\mu = 3{\cdot}9863\ldots \times 10^{20} \text{ cm}^3/\text{sec}^2.$$

This value agrees well with astronomical determinations, though these are possibly less reliable at the present time since they necessarily involve a knowledge of some planetary or asteroidal distance in metric units. A more accurate value should eventually be obtained from artificial satellites.

EXERCISES

1. Show, from the conservation integrals (§1.10), that the extremal distances between a gravitating pair of masses μ_1 and μ_2 in elliptical orbit are the roots of

$$Ed^2 + \mu_1\mu_2 d - \frac{\mu_1 + \mu_2}{2\mu_1\mu_2} h^2 = 0.$$

Obtain the corresponding geometrical connexion

$$d^2 - 2ad + al = 0$$

and hence re-derive the relations between E, h and the orbital elements a, l.

2. A particle at the equator is released from rest relative to the earth and falls freely. Show that the eccentricity of its orbit about the earth's centre in a sidereal frame is $1 - a\Omega^2/g$ very nearly, where a is the equatorial radius, g is mean gravity at sea-level, and Ω is the earth's spin [cf. the second of (11) in §1.7].

3. A plane orbit is observed from various frames of reference rotating about a perpendicular axis through an origin G in the plane. In frame F_0 the apparent orbit is central with respect to G, with inward acceleration $f_0(r)$ and constant rate of sweep η_0. Show that the apparent orbit in another frame F is also central, with acceleration $f(r)$ and rate of sweep η, where

$$f(r) - f_0(r) = 4(\eta^2 - \eta_0^2)/r^3, \qquad \eta/\eta_0 = k,$$

provided that F rotates relative to F_0 in such a way that the respective inclinations θ and θ_0 of the radius vector maintain a fixed ratio $\theta/\theta_0 = k$ (*Principia:* Book I, Props. XLIII and XLIV). Verify that the total energy of an orbiting mass-point appears to have the same constant value in all frames such as F, independently of k.

In particular, when the orbit $r(\theta_0)$ in F_0 is an ellipse with semi-latus rectum l, show that the acceleration in F is

$$f(r) = \frac{\mu}{r^2}\left(1 + \frac{\lambda}{r}\right) \text{ where } \mu = \frac{4\eta_0^2}{l}, \qquad \lambda = (k^2 - 1)l.$$

Verify that the apsidal angle of the orbit in F is πk, and hence that an apse advances through the angle $2\pi(k - 1)$ in each libration (i.e. while the radius varies from one extreme value to the other and back again).

4. An ellipse with major axis $2a$ and eccentricity e is described under an acceleration μ/r^2 to its focus.

(i) Show that the outward radial component of velocity is

$$u = \frac{\mu e}{2\eta} \sin\theta$$

where

$$r(1 + e\cos\theta) = a(1 - e^2) = 4\eta^2/\mu,$$

η being the rate of sweep of area. Conversely, suppose that the velocity vector is initially given at a certain radius, or equivalently the constant η with the starting values of r and u. Explain how this initial data determines the orbital elements a, e, together with the relative inclination θ of the line of apses, and obtain as elegant a system of inverse formulae as you can.

(ii) Show that the effects of an infinitesimal change in the initial value of u, but not in η or r, are given by

$$\frac{\delta e}{\sin\theta} = \frac{(1 - e^2)}{2e\sin\theta} \cdot \frac{\delta a}{a} = \frac{e\,\delta\theta}{\cos\theta} = \frac{2\eta\,\delta u}{\mu}.$$

(iii) A bounded but unclosed orbit, in which the congruent successive loops are near-ellipses (a, e), is described under the purely radial acceleration

$$\frac{\mu}{r^2}\left(1 + \frac{\lambda}{r^n}\right), \qquad \lambda \text{ infinitesimal.}$$

By regarding the last equality in (ii) as defining an 'influence function', deduce that an apse line advances through the angle

$$\frac{2\lambda}{el^n}\int_0^\pi (1 + e\cos\theta)^n \cos\theta\, d\theta$$

per revolution (or libration), where $2l$ is the latus rectum. [As Newton himself emphasized, the virtual absence of such secular (or progressive) rotations of planetary orbits, over long periods of time, is a sensitive test of the inverse-square law of gravitation.]

5. Use the final formula in the preceding question to treat the following situations in a uniform manner.

(i) In Newton's theory of revolving orbits $n = 1$. Show that the apsidal advance is $\pi\lambda/l$ per revolution, in agreement with the exact formula in question (3) when λ is small and k near unity.

(ii) When $n = 2$ show that the apsidal advance is $2\pi\lambda/l^2$ in each revolution. [In Einstein's relativistic theory of planetary orbits $\lambda = 3\mu l/c^2$ where c is the speed of light and μ is the solar mass. For an artificial satellite orbiting in the earth's equatorial plane $\lambda = (\varepsilon - \beta)r_0^2$ where r_0 is the earth's mean radius and $\varepsilon - \beta$ depends on its oblateness, etc., see §1.18, equation (10).]

(iii) Suppose that a tenuous cloud of dust with uniform gravitational mass-density σ surrounds the sun, so that a planet moves through it freely but experiences an additional heliocentric acceleration $\frac{4}{3}\pi\sigma r$ [cf. §1.16, equation (1)]. Assuming the eccentricity to be small, show that the perihelion retrogresses at rate $\sigma\tau$, where τ is the orbital period.

6. The earth, moon, and sun are regarded as an isolated system. E, M, S denote their gravitational masses, $\mathbf{r}$ the vector distance between earth and moon, and $\mathbf{R}$ the vector distance between the E-M mass-centre and the overall mass-centre G. Show that the constant overall moment of momentum about G (§1.10) is expressible as

$$\left(1 + \frac{E + M}{S}\right)(E + M)\mathbf{R} \wedge \dot{\mathbf{R}} + \frac{EM}{E + M}\mathbf{r} \wedge \dot{\mathbf{r}}$$

and interpret the various terms.

From the equations of motion of M and E (§1.6) show that

$$\frac{(E + M)}{S}\frac{d}{dt}(\mathbf{R} \wedge \dot{\mathbf{R}}) = \frac{2EM}{(E + M)}\left(\frac{1}{r_M{}^3} - \frac{1}{r_E{}^3}\right)\Delta$$

where Δ is the current area of the triangle configuration and r_M, r_E the distances of M and E from S.

7. Establish the following special solution of the 3-body problem (a generalization of §1.8, also due to Lagrange). Gravitating masses μ_1, μ_2, μ_3 move in a plane of fixed orientation in a sidereal frame, so that they form an equilateral triangle with varying side $a(t)$ and spin $\Omega(t)$. By considering radial and transverse accelerations with respect to G in the presumed motion, show that

$$\ddot{r}_j - r_j\Omega^2 = -(\mu_1 + \mu_2 + \mu_3)r_j/a^3,$$
$$r_j{}^2\Omega = \text{constant} \qquad (j = 1, 2, 3),$$

with the notation of §1.8. Verify that these six equations are all satisfied provided

$$\ddot{a} - a\Omega^2 = -(\mu_1 + \mu_2 + \mu_3)/a^2, \qquad a^2\Omega = \text{constant}.$$

Prove that the motion can be initiated by projecting the masses from an equilateral configuration with speeds proportional to the radii from G and in directions equally inclined to them.

Why are the orbits about G similar ellipses and how does the motion appear in a frame rotating with spin Ω?

8. In the restricted problem of 3 bodies (§1.9) examine the character of the U-surface near the five points of libration, through the second-order terms in the local Taylor expansions. For this purpose obtain the following derivatives at a general position (ξ, η):

$$\frac{\partial^2 U}{\partial \xi^2} = \frac{\mu_1 + \mu_2}{a^3} + \frac{\mu_1}{d_1^5}(2x_1^2 - \eta^2) + \frac{\mu_2}{d_2^5}(2x_2^2 - \eta^2),$$

$$\frac{\partial^2 U}{\partial \eta^2} = \frac{\mu_1 + \mu_2}{a^3} + \frac{\mu_1}{d_1^5}(2\eta^2 - x_1^2) + \frac{\mu_2}{d_2^5}(2\eta^2 - x_2^2),$$

$$\frac{\partial^2 U}{\partial \xi \partial \eta} = 3\eta \left(\frac{\mu_1 x_1}{d_1^5} + \frac{\mu_2 x_2}{d_2^5}\right),$$

where $x_1 - x_2 = a$, $\mu_1 x_1 + \mu_2 x_2 = (\mu_1 + \mu_2)\xi$. Hence, or otherwise, show that the second-order increment in $aU(\xi, \eta)$ in a displacement from either triangular point of libration is

$$\tfrac{3}{2}[\mu_1(\delta\varepsilon_1)^2 + \mu_2(\delta\varepsilon_2)^2]$$

where $\delta\varepsilon_1$ and $\delta\varepsilon_2$ are the fractional changes in length of the sides of the triangle.

Show that in a displacement from a collinear point of libration the increment is

$$\tfrac{3}{2}(\mu_1 + \mu_2)\left(\frac{\delta\xi}{a}\right)^2 + \frac{\mu_1}{\lambda - 1}\left(1 - \frac{1}{|\lambda|^3}\right)\left[\left(\frac{\delta\xi}{a}\right)^2 - \frac{1}{2}\left(\frac{\delta\eta}{a}\right)^2\right],$$

where λ is the appropriate root of (3) in §1.8. Establish that such positions are cols by proving that λ is contained in the interval $(-1, 2)$ whatever the mass ratio.

Verify that at the triangular positions

$$U = \frac{3(\mu_1 + \mu_2)}{2a}\left[1 - \frac{\mu_1 \mu_2}{3(\mu_1 + \mu_2)^2}\right]$$

is an absolute (and not merely local) minimum.

9. If the law of gravitation were inverse cube instead of inverse square, show that Lagrange's identity for an n-body system, (3) in §1.11, would become $\ddot{T} = -4E$. Deduce that the system would either coalesce in a finite time or would disperse, at least in part, as $t \to \infty$. Discuss the possibilities in detail.

10. Investigate the differential field of acceleration in the neighbourhood of an origin O due to a mass-point μ in position $\mathbf{r} \equiv r\mathbf{l}$ relative to O.

(i) In terms of spherical polar coordinates (ρ, θ, ψ) with respect to O and the axis $\mathbf{l}$ (cf. §1.16), show that the Legendre expansion of the gravitational potential of μ near O in powers of ρ/r is

$$\frac{-\mu}{r}\left[1 + \frac{\rho}{r}P_1(\theta) + \frac{\rho^2}{r^2}P_2(\theta) + \ldots\right].$$

(ii) Identify the P_1 term with the potential of a uniform field of acceleration $\mu\mathbf{l}/r^2$ and, by using (3) in §1.16 or otherwise, obtain the remaining differential acceleration as

$$\frac{\mu}{r^2}\left[\frac{3(\mathbf{l}\boldsymbol{\rho})\mathbf{l} - \boldsymbol{\rho}}{r} + \ldots\right].$$

(iii) Show that the P_2 contribution exhibited here is equivalent to a central acceleration $2\mu\boldsymbol{\rho}/r^3$ outwards from O together with a centripetal acceleration towards the axis $\mathbf{l}$, in amount $3\mu/r^3 \times$ distance from the axis; or alternatively to the resultant field due to a mass $\frac{1}{2}\mu$ at $\mathbf{r}$ and a fictitious 'image' mass $\frac{1}{2}\mu$ at $-\mathbf{r}$ (to the same order of approximation).†

† When O is identified with the earth's centre and μ with the moon or sun, the analysis furnishes the luni-solar 'tide-generating' field. In a state of equilibrium under the component (iii) a gravitating mass of uniform fluid becomes a prolate spheroid with axis $\mathbf{l}$, by analogy with the oblate spheroid possible under the centrifugal field associated with axial spin (§1.18). This idea is pursued in the *Principia* (Book III, Prop. XXXVI *et seq.*) and, with due modifications, in modern theories of the semi-diurnal tide. See, for instance, G. H. Darwin: *The Tides* (John Murray, London 1911, 3rd edition; Golden Gate Edition 1962).

Also, when O is the centre of any planet and μ is the sun, (ii) furnishes the solar 'disturbing field' perturbing the orbit of a satellite. Compare the discussions of solar action on the earth–moon system in the *Principia* (Book III, Prop. XXV *et seq.*) and in V. M. Blanco and S. W. McCuskey: *Basic Physics of the Solar System*, Chapter V (Addison-Wesley 1961) or R. Kurth: *Introduction to the Mechanics of the Solar System*, Chapter III, §6 (Pergamon Press 1959).

CHAPTER II

FOUNDATIONS OF MECHANICS: GENERAL PRINCIPLES

2.1 Frames of Reference and Rates of Change

In the preceding account of celestial mechanics the frame of reference was generally sidereal, and in particular Newtonian, since the basic equations for an arbitrary system are thereby reduced to their simplest form. Exceptionally, in the restricted problem of three bodies, a frame revolving with the primary masses was shown to have certain advantages. In terrestrial mechanics, on the other hand, it is clearly convenient to adopt a 'natural' frame of reference, rigidly attached to the earth or at least moving relatively to it in an assigned way. Yet again, in space travel the natural frame is provided by the vehicle, while this itself is navigated by reference to the stars or the nearest planets. We need, therefore, to be able to relate the apparent motions of a body or a system seen from different frames. This analysis is moreover indispensable subsequently in clarifying the concept of force.

Suppose that the path of a moving point is observed from some particular frame, denoted by α ('acting' or 'adopted'). From the apparent velocity and acceleration in α it is required to determine their apparent values in some other frame, perhaps more fundamental, denoted by β ('background' or 'basic'). For example, α might be terrestrial and β sidereal. The relative motion of the two frames themselves will usually have both translational and rotational components. For the present it is enough to suppose that the translation is arbitrary but that the axis of rotation has a constant direction with respect to either frame.† Select orthogonal triads of right-handed axes (ξ, η, ζ) and (ξ', η', ζ') fixed in α and β respectively, and for convenience let ζ and ζ' be parallel to the axis of rotation. The spin of α with respect to β is then $\dot{\psi}$ where ψ is the angle between ξ' and ξ or η' and η (Fig. 9).

For the sake of applications in other contexts we find the rate of

† For the general analysis when the axis of rotation varies in direction see §3.1.

change of an arbitrary vector quantity and then particularize. At a generic instant let any free line-segment representing the quantity (in magnitude and direction only) have orthogonal projections (p, q, r) and (p', q', r') on the respective axes. These will be said to be the components 'referred to', or simply 'on', the axes; by their

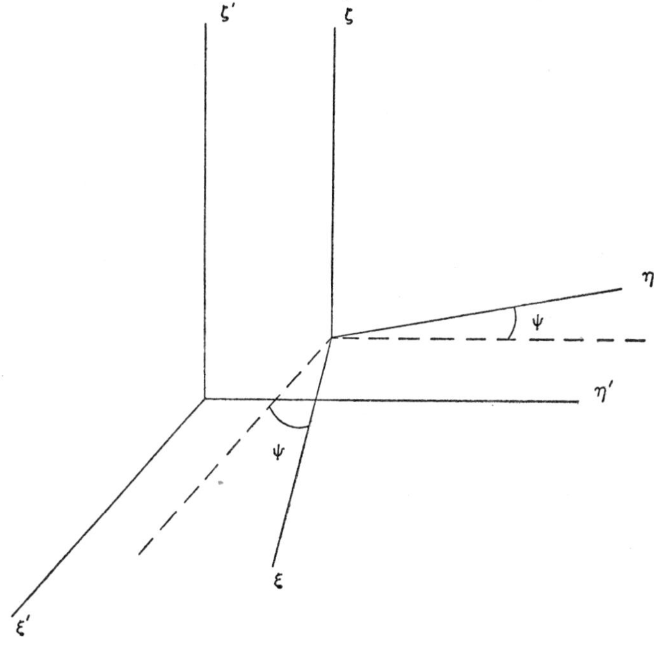

FIG. 9

manner of definition, such components have primarily a geometric significance. The components $(\dot{p}, \dot{q}, \dot{r})$ on the axes in α, where the dot denotes the time derivative, define a vector which will be called the rate of change of the considered quantity relative to α. (Its components on other axes† in α are always the time-derivatives of the corresponding components of the quantity.) Similarly, $(\dot{p}', \dot{q}', \dot{r}')$ on the axes in β form the vector rate of change relative to β. Since our aim is to find a connexion between these apparent rates we must compare their components on the *same* axes, say those in the adopted frame α for preference.

† The remarks in §1.2 are recalled.

We begin with the equations of coordinate transformation:

$$\begin{pmatrix} p' \\ q' \end{pmatrix} = p \begin{pmatrix} \cos \psi \\ \sin \psi \end{pmatrix} + q \begin{pmatrix} -\sin \psi \\ \cos \psi \end{pmatrix}$$

together with $r' = r$. These are written here in a form exhibiting the (ξ', η') components of unit vectors attached to the (ξ, η) axes. By differentiating these equations and grouping terms to exhibit the same unit vectors:

$$\begin{pmatrix} \dot{p}' \\ \dot{q}' \end{pmatrix} = (\dot{p} - q\dot{\psi}) \begin{pmatrix} \cos \psi \\ \sin \psi \end{pmatrix} + (\dot{q} + p\dot{\psi}) \begin{pmatrix} -\sin \psi \\ \cos \psi \end{pmatrix}$$

together with $\dot{r}' = \dot{r}$. Consequently, the rate of change relative to β has the components $(\dot{p} - q\dot{\psi}, \dot{q} + p\dot{\psi}, \dot{r})$ on the axes in α.

This simple but fundamental result can be stated in the following way. Referred to axes in an adopted frame α, a vector quantity (p, q, r) has a rate of change $(\dot{p}, \dot{q}, \dot{r})$ relative to α, but a rate of change

$$(\dot{p} - q\Omega, \dot{q} + p\Omega, \dot{r}) \tag{1}$$

relative to any other frame β, where $(0, 0, \Omega)$ is the spin of α relative to β. Notice that the formula in no way involves particular axes in the background frame β, and does not depend on the relative translation of the two frames but only on their relative spin. The difference between the apparent rates, $(-q, p, 0)\Omega$, is clearly interpretable invariantly as the rate of change relative to β of a segment of fixed orientation in α when momentarily parallel to the considered vector itself.

The roles of the frames are naturally interchangeable and the theorem can be formulated symmetrically: the difference in the apparent rates of change depends only on the spin and on the current value of the vector quantity. This statement implies that the precise dependence is recoverable by evaluating this difference in respect of (for convenience) a fixed segment in either frame. A far-reaching corollary should be well noted: any instantaneous addition to the rate of change without alteration in the quantity itself, appears the same in all frames. This remains true when the spin axis varies in direction (§3.1).

By taking as the considered quantity the vector rate of change itself in β, (1) furnishes the components on the axes in α of the 'second rate of change' relative to β:

$$\left.\begin{aligned} \frac{d}{dt}(\dot{p} - q\Omega) - (\dot{q} + p\Omega)\Omega &= \ddot{p} - \Omega^2 p - 2\Omega\dot{q} - \dot{\Omega}q, \\ \frac{d}{dt}(\dot{q} + p\Omega) + (\dot{p} - q\Omega)\Omega &= \ddot{q} - \Omega^2 q + 2\Omega\dot{p} + \dot{\Omega}p, \\ \frac{d}{dt}(\dot{r}) &= \ddot{r}. \end{aligned}\right\} \tag{2}$$

The second rate of change relative to α has of course corresponding components $(\ddot{p}, \ddot{q}, \ddot{r})$ simply. By repeating this procedure, expressions can be generated for a rate of change of any order, and it is clear by induction that the difference between values for the two frames depends solely on rates of change of all lower orders, together with the spin and its derivatives of similar orders. We thus have the general theorem: any instantaneous addition to a rate of change of order n, unaccompanied by alterations in rates with orders up to $n - 1$, is invariant with respect to frames of reference.

We return to our initial task, and for (p, q, r) substitute the coordinates (ξ, η, ζ) of the position vector of a moving point with respect to the origin of axes in α. From (1) the velocity relative to this origin as seen from β has components

$$(\dot{\xi} - \Omega\eta,\ \dot{\eta} + \Omega\xi,\ \dot{\zeta}) \tag{3}$$

on the axes in α, whereas the velocity seen from α has components $(\dot{\xi}, \dot{\eta}, \dot{\zeta})$. From (2) the acceleration relative to the origin in α as seen from β has components

$$\left.\begin{aligned} &\ddot{\xi} - \Omega^2\xi - \dot{\Omega}\eta - 2\Omega\dot{\eta}, \\ &\ddot{\eta} - \Omega^2\eta + \dot{\Omega}\xi + 2\Omega\dot{\xi}, \\ &\ddot{\zeta}, \end{aligned}\right\} \tag{4}$$

on the axes in α, whereas the acceleration seen from α has components $(\ddot{\xi}, \ddot{\eta}, \ddot{\zeta})$. The extra terms form the following vectors:†

† The invariant vectorial expressions are derived in full generality in §3.1. In an introductory account greater clarity is obtained by displaying cartesian components of all rates of change, and especially of accelerations.

(i) $-\Omega^2(\xi, \eta, 0)$, centripetally towards the spin axis placed at the considered origin in α;

(ii) $\dot{\Omega}(-\eta, \xi, 0)$, perpendicular to the centripetal radius;

(iii) $2\Omega(-\dot{\eta}, \dot{\xi}, 0)$, perpendicular to the apparent path in α.

All are transverse to the axis of rotation and are unaffected by the component motion parallel to this axis. Term (iii) is referred to as the Coriolis acceleration, after the French engineer who first derived it (1829). It is evidently proportional to the speed in the transverse component of the *apparent motion in* α, and is absent when this motion vanishes or is parallel to the axis of rotation. Consequently, the other terms in (4) collectively represent the acceleration in β, and with respect to the α origin, of the point (ξ, η, ζ) when *fixed* in α.

Note carefully that Ω is the spin of α, not the spin of the centripetal radius (whose orientation in α usually varies). These spins coincide only when the apparent motion in α happens to be confined to some fixed plane, $\xi/\eta =$ constant. Suppose, without loss of generality, that the point remains on the ξ axis so that $\eta = 0$. Expressions (4) then reduce to $(\ddot{\xi} - \Omega^2\xi, \dot{\Omega}\xi + 2\Omega\dot{\xi}, 0)$, the familiar polar components of acceleration in β. In this derivation the frame α plays only an auxiliary role.

Finally, we draw from the general invariance theorem a basic principle in kinematics. If, at some instant, there is a *finite change* in acceleration (of a mass-point say), without an accompanying change in position or velocity at that instant, this *additional* acceleration appears to be the same in all frames of reference (whereas the *total* acceleration of course does not). A similar statement holds in respect of the *difference* between the acceleration as it actually is and as it would be, supposing that the dynamical situation were otherwise, provided that the imagined comparison is made when the mass-point has the same position and velocity.

2.2 Motion Relative to the Earth

Terrestrial phenomena are commonly described in terms of the frame provided by the earth itself. Over any ordinary interval of time, short compared with the period of equinoctial precession, the earth's spin Ω can be regarded as constant in direction relative to a sidereal frame, as well as constant in magnitude. The preceding analysis then applies, with $\dot{\Omega} = 0$, and the (ξ, η, ζ) axes fixed in the earth can conveniently be chosen with origin at its centre G and ζ northward along the polar axis (Fig. 10).

Suppose that we require, for a particle P which is not necessarily isolated or *in vacuo*, the (ξ, η, ζ) components of acceleration $\mathbf{a}$ with respect to G and relative to any sidereal frame. These are given by (4) in §2.1 with $\dot{\Omega} = 0$. But, by (6) in §1.18, the centripetal contribution from Ω is precisely $\mathbf{f} - \mathbf{g}$, the difference at P between the

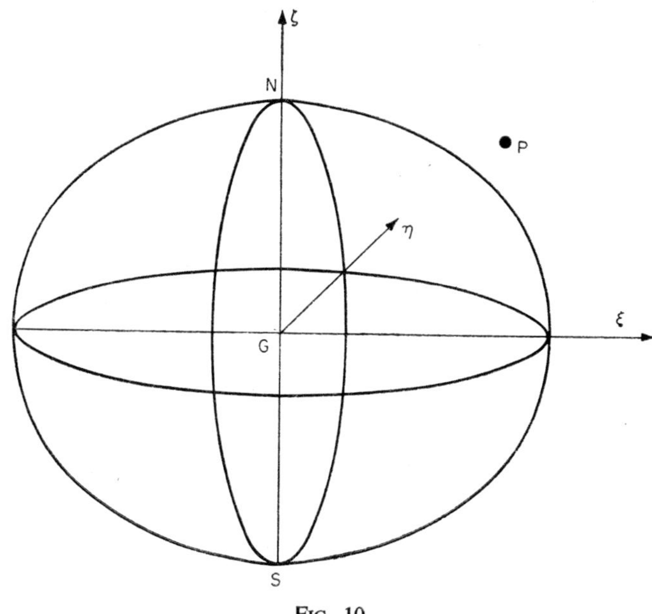

FIG. 10

earth's field and gravity (defined as the negative gradient of the geopotential). Thus

$$\mathbf{a} - \mathbf{f} + \mathbf{g} = (\ddot{\xi}, \ddot{\eta}, \ddot{\zeta}) + 2\Omega(-\dot{\eta}, \dot{\xi}, 0) \tag{1}$$

when the centripetal contribution is transferred to the other side of the equation for convenience.

If, now, P is a free particle we know from general gravitation theory (§1.6) that $\mathbf{a}$ is just the earth's field $\mathbf{f}$, provided that the very small difference between the accelerations of P and G due to other matter is ignored.† Then, on the left side of (1) there remains only **g**, and we notice that the apparent acceleration $(\ddot{\xi}, \ddot{\eta}, \ddot{\zeta})$ deviates from it in magnitude and direction on account of the Coriolis effect of the earth's rotation. Only when P is momentarily at rest in the earth's frame (but necessarily moving in a sidereal frame) is the acceleration equal to local gravity. This establishes the qualification anticipated in §1.18.

Suppose, more generally, that P is isolated but not free, and that the difference $\mathbf{a} - \mathbf{f}$ is a known function of various kinematic variables. For example, according to experiment, the additional retardation due to the earth's atmosphere depends essentially on height and relative velocity, for a given mass; or again, the additional acceleration produced by certain modes of propulsion is a predetermined function of time. Then, with initial conditions on position and velocity, equations (1) suffice to determine the path of P as it appears from the terrestrial frame. This viewpoint is particularly convenient when the apparent path has only a limited range; otherwise the sidereal frame is preferable, for instance in treating the motion of an artificial satellite or a ballistic missile. In principle the gravity vector is a function of the coordinates that is known from experiment, or at any rate from theory via (4) and (5) in §1.18. The differential equations are therefore non-linear even through this dependence alone, and in general can only be solved by numerical methods.

However, when the range of the apparent path is small compared with the earth's radius, in ratio not more than order 10^{-3} say, the spatial variation of gravity and the earth's field can be neglected. The effect of this approximation is to linearize the equations for a free particle. (Equivalently, the path in a sidereal frame is treated as a parabolic arc, the well-known orbit in a uniform field of acceleration.) Let g denote the magnitude of this mean gravity through the region traversed by the path, and λ the astronomical latitude defining the inclination of mean gravity to the plane of the equator; when the path happens to lie close to the earth's surface λ can be

† It is easy to estimate this difference from the inverse-square law: the contribution from a celestial body with a mass of n earth-units at a distance of d earth-radii is at most $2ng/d^3$ (very nearly). For the moon this is about $10^{-7}g$ and for the sun still less by a factor of two or so. Relatively minute though the differential luni-solar field is in the present context, however, it is responsible for the oceanic tides.

taken as ordinary geographic latitude, well within the accuracy attempted. Equations (1), specialized for free orbit, are then

$$\left.\begin{aligned}\ddot{\xi} - 2\Omega\dot{\eta} &= -\,g\cos\lambda,\\ \ddot{\eta} + 2\Omega\dot{\xi} &= 0\qquad\quad,\\ \ddot{\zeta} \qquad\quad &= -\,g\sin\lambda,\end{aligned}\right\} \tag{2}$$

when the (ξ, ζ) plane is taken as the meridian containing the direction of mean gravity. Notice that the component motions parallel and transversely to the polar axis are 'uncoupled'; that is, the particle 'falls' with constant acceleration $g \sin \lambda$ towards the equatorial plane while its transverse motion is executed independently. The integration of this set of linear second-order equations is correspondingly simple. With values at time $t = 0$ indicated by a subscript zero,

$$\left.\begin{aligned}\xi - \xi_0 - \dot{\xi}_0 t &= \dot{\xi}_0\left(\frac{\sin 2\Omega t}{2\Omega} - t\right) + \left(\dot{\eta}_0 - \frac{g\cos\lambda}{2\Omega}\right)\left(\frac{1-\cos 2\Omega t}{2\Omega}\right),\\ \eta - \eta_0 - \dot{\eta}_0 t &= \left(\dot{\eta}_0 - \frac{g\cos\lambda}{2\Omega}\right)\left(\frac{\sin 2\Omega t}{2\Omega} - t\right) - \dot{\xi}_0\left(\frac{1-\cos 2\Omega t}{2\Omega}\right),\\ \zeta - \zeta_0 - \dot{\zeta}_0 t &= -\tfrac{1}{2}gt^2\sin\lambda.\end{aligned}\right\} \tag{3}$$

The (ξ, η) solution can be seen as the superposition of a uniform translation in the η direction with speed $g \cos \lambda/2\Omega$ and a uniform circular motion at angular rate 2Ω in a radius

$$\left[\dot{\xi}_0^2 + \left(\dot{\eta}_0 - \frac{g\cos\lambda}{2\Omega}\right)^2\right]^{\frac{1}{2}} \Big/ 2\Omega.$$

This solution remains realistic only during a time interval consistent with the linearization, which requires at least that $\frac{1}{2}gt^2$ should not exceed about $10^{-3}\times$ earth's radius, or that t should be less than 30 seconds say. Then Ωt, the earth's rotation, is less than 2×10^{-3} radians and so the periodic terms in (3) can very well be replaced by cut-off expansions in the variable Ωt:

$$\left.\begin{aligned}\xi - \xi_0 - \dot{\xi}_0 t &= -\tfrac{1}{2}gt^2\cos\lambda + \Omega\dot{\eta}_0 t^2 + \ldots,\\ \eta - \eta_0 - \dot{\eta}_0 t &= \Omega(\tfrac{1}{3}gt\cos\lambda - \dot{\xi}_0)t^2 + \ldots,\\ \zeta - \zeta_0 - \dot{\zeta}_0 t &= -\tfrac{1}{2}gt^2\sin\lambda,\end{aligned}\right\} \tag{4}$$

where the omitted remainders have Ω^2 as a factor.

The terms without Ω constitute the familiar parabolic orbit in a fixed vertical plane, when only the centripetal effects of the earth's rotation are allowed for (via **g**). The extra terms in t^2 are evidently interpretable as displacements due to a uniform field of acceleration $2\Omega(\dot{\eta}_0, -\dot{\xi}_0)$, associated with the Coriolis effect at $t = 0$. The extra term in t^3 compensates for the change in the transverse motion; to the present order of approximation this is produced by the gravity component $g \cos \lambda$. In fact, it is the displacement due to a varying apparent field $2\Omega t(0, g \cos \lambda)$ and is recoverable from this by two-fold integration.

To assess the quantitative contribution by the latter, consider in particular free fall from rest relative to the earth: $\dot{\xi}_0 = \dot{\eta}_0 = \dot{\zeta}_0 = 0$. The descent is not precisely vertical but deviates to the east by $\frac{1}{3}\Omega g t^3 \cos \lambda$. (From the sidereal viewpoint the effect is immediately understandable, since the initial speed in *that* frame is $\Omega\xi$ eastward, in excess of the speeds of points beneath which turn with the earth.) In a time of 10 seconds, say, corresponding to a fall of some 500 metres, the deviation at latitude 45° is about 15 cm only; nevertheless, even so small an effect was observed as long ago as 1833 in experiments in accurately surveyed mineshafts.† The neglected remainders in Ω^2 are utterly insignificant, being less by several orders of magnitude. The contribution $\Omega(\dot{\eta}_0, -\dot{\xi}_0)t^2$ is comparable only when the initial transverse speed is high; at 10^5 cm/sec, which is of ballistic order, the amount is about 7 metres in the same interval (the horizontal range being still within the permitted limits). On the other hand, when the time interval is prolonged, even though the speed remains moderate, the cumulative Coriolis effects can be highly significant; of course, in such a combination of circumstances the particle could not be free and the full equations (1) would have to be solved, without linearization. Examples of such large-scale phenomena are the geostrophic wind and the deviation of oceanic currents.

2.3 Motion under Independent Constraint

So far our concern has been exclusively with isolated bodies in space, and with their linear motions in conformity with the gravitational postulates. We turn now to motion of a particle in a gravitational field under a superimposed geometric constraint that is

† R. Dugas: *A History of Mechanics*, p. 380 (Routledge and Kegan Paul 1957).

pre-determined in space and time. In other words, it is a constraint that operates entirely *independently* of the behaviour or the mass of the particle. Specifically, we consider types of constraint such that the permitted paths $\mathbf{r}(t)$ are subject at all times to the overriding restriction

$$\psi(\mathbf{r}(t), t) \equiv 0 \tag{1}$$

where $\psi(\mathbf{r}, t)$ is a given scalar function of the coordinates (in a stated frame of reference) and perhaps also of the time. At each instant such an equation represents either a real surface in physical contact with the particle (such as the wall of a containing vessel) or the locus of a point of attachment of the mass (such as the end of an inextensible but flexible cord). Whether actual or not, the surface is fixed in relation to the stated frame when its equation is $\psi(\mathbf{r}) = 0$; in general, t enters explicitly and the surface either moves rigidly or with stipulated deformation. Two different relations (1) may subsist simultaneously, in which case the particle must move so as to be currently situated somewhere on the instantaneous curve of intersection; if neither relation involves t the actual orbit coincides with a pre-determined curve fixed in the particular frame of reference.

Since the function of time on the left of (1) is identically zero throughout a specifiable interval, its derivatives of all orders also vanish. In particular, when ψ has continuous partial derivatives at the considered position and time,

$$(\text{grad } \psi)\frac{d}{dt}\mathbf{r}(t) + \frac{\partial \psi}{\partial t} = 0.$$

Thus the velocity $d\mathbf{r}/dt$ of the particle appears in scalar product with grad ψ, which is orthogonal to the current surface of constraint. Consequently, the normal component of velocity, and only this component, is obtainable as a pre-determined function of position and time, expressible in terms of the partial derivatives of $\psi(\mathbf{r}, t)$. In fact, the normal velocity is equal to the local speed of advance of the surface itself, since the previous equation naturally applies also to a constrained point whose orbit is an orthogonal trajectory of the family of surfaces. A further differentiation yields the acceleration $d^2\mathbf{r}/dt^2$ in scalar product with grad ψ, the other terms involving only the components of velocity and the second partial derivatives of ψ. Thus, for a constraint of type (1) whether varying or not, the normal component of acceleration, and only this component, is a pre-determined function of position, velocity, and time.† And so on for accelerations of all higher orders.

It remains to state principles governing the tangential motion. For the present we restrict attention to constraints that are *frictionless*, in the sense that the additional accelerations associated with them have no tangential component. That is to say, any particle on which such a constraint operates has the tangential acceleration it would have in the absence of the constraint, if other circumstances were *arbitrary but unchanged* (such as position, velocity, and gravitational field). This is a matter of pure definition, whose self-consistency is guaranteed by the previous theorem; it is for experimental physics to determine which constraints in nature actually have this property. Expressed formally, the *additional* acceleration due to the constraint is $c\mathbf{n}$, where $\mathbf{n}(\mathbf{r}, t)$ is the local unit normal to the surface $\psi(\mathbf{r}, t)$ and $c(t)$ is some function of time depending on the particular motion. By the invariance theorem of §2.1 c is the same in all frames of reference. The constraint is said to be *one-sided* when it can only operate in such a way that c has a pre-determined sign.

When the particle is otherwise free to move in the gravitational field near the earth's surface,

$$\mathbf{a} - \mathbf{f} = c\mathbf{n} \tag{2}$$

with the notation of §2.2. From (1) there, writing $\mathbf{v} \equiv (\dot{\xi}, \dot{\eta}, \dot{\zeta})$ for the velocity relative to the earth and $\mathbf{\Omega} \equiv (0, 0, \Omega)$, we then have

$$\dot{\mathbf{v}} + 2\mathbf{\Omega} \wedge \mathbf{v} = \mathbf{g} + c\mathbf{n} \tag{3}$$

where $\dot{\mathbf{v}}$ is the acceleration relative to the earth. Together with the condition of constraint these equations are sufficient to determine the path and the function c.

Suppose, now, that the constraint surface is fixed to the earth. Then $\mathbf{nv} = 0$ at all times since the velocity $\mathbf{v}$ is purely tangential, and hence $d(\mathbf{nv})/dt = 0$. The normal component of acceleration $\mathbf{n}d\mathbf{v}/dt$ can therefore be calculated as $-\mathbf{v}d\mathbf{n}/dt$, following the path. But, by a basic formula in the differential geometry of surfaces,

$$d\mathbf{n}d\mathbf{s} = -\kappa(d\mathbf{s})^2$$

† This result might almost be regarded as self-evident. But, as with other such theorems in pure kinematics, intuition should be grounded on a precise proof from first principles.

where $d\mathbf{s}$ is any infinitesimal arc in the surface. $d\mathbf{n}$ is the corresponding increment in the unit normal to the surface (and is tangential since $\mathbf{n}^2 = 1$ and $\mathbf{n}d\mathbf{n} = 0$, though not necessarily in direction $d\mathbf{s}$). κ is the curvature of the normal section of the surface through the path,† and is positive or negative according to whether $\mathbf{n}$ is directed towards the concave or convex side of the section. Whence, since $\mathbf{v} = d\mathbf{s}/dt$ along the actual path,

$$\mathbf{n}\dot{\mathbf{v}} = \kappa\mathbf{v}^2,$$

which expresses the normal acceleration solely in terms of position and velocity, in illustration of the opening theorem. Then, from (3),

$$c = \kappa\mathbf{v}^2 + \mathbf{n}(2\boldsymbol{\Omega} \wedge \mathbf{v} - \mathbf{g}). \tag{4}$$

Still assuming the constraint fixed to the earth, we notice an immediate first integral of (3), through the usual procedure of forming its scalar product with $\mathbf{v}$ (cf. §1.9). This annihilates the terms in $\boldsymbol{\Omega}$ and $\mathbf{n}$, leaving

$$\mathbf{v}(\dot{\mathbf{v}} + \text{grad } \Phi) = 0 \quad \text{where } \mathbf{g} = -\text{ grad } \Phi,$$

and Φ is the geopotential (§1.18). Whence, integrating with respect to time,

$$\tfrac{1}{2}\mathbf{v}^2 + \Phi = \text{constant}, \tag{5}$$

since Φ is a function only of position in this frame. When the gravity field can be treated as approximately uniform,

$$\mathbf{g}/g \equiv -\text{ grad } z$$

where z is a local coordinate along the upward vertical, and then

$$\Phi = gz \tag{6}$$

as ordinarily assumed.

Lastly, it is worth noting that constraints with Coulomb (sliding) friction can readily be brought within this conceptual framework: it is only necessary to re-phrase the conventional definition. The additional acceleration due to the constraint is directed outwards in the normal plane through the *relative* motion, and is inclined to

† κ must not be confused with the principal curvature of the path, which is $\kappa \sec \theta$ where θ is the angle between the osculating plane and the plane of normal section (Meusnier's theorem).

this motion at a fixed angle $\frac{1}{2}\pi + \varepsilon$ where ε is the so-called angle of friction and is acute.

2.4 Foucault Pendulum Experiment

An elementary form of frictionless constraint is (ideally) provided by simple pendular support *in vacuo*. In principle this term covers any means by which a particle is maintained at a fixed distance l from some point O whose motion is specifiable. The surface of

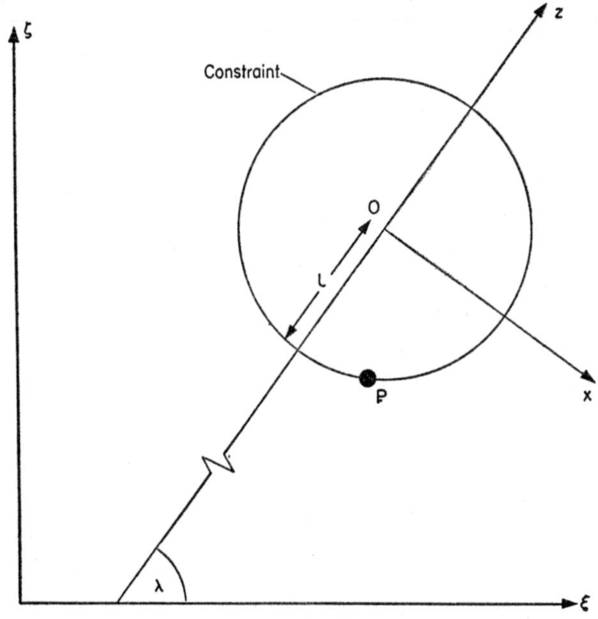

FIG. 11

constraint is then a sphere of radius l with O as centre. For instance, the particle might be in contact with an actual spherical container or attached to a taut but flexible string. In what follows, it is supposed that the surface is fixed with respect to the earth.

We begin by remarking that the particle can itself remain at rest relative to the earth. In fact, possible positions of relative equilibrium are the two points where local gravity is normal to the spherical surface (whether the positions are actual or not depends on the mechanism). This follows since at such a point the particle would have no tangential acceleration in the terrestrial frame if the constraint were absent, and therefore remains at rest if placed there without relative velocity when the constraint operates.

Arbitrary motion in a neighbourhood of the lower position of equilibrium is now examined. For this purpose it is convenient to introduce new axes (x, y, z) with origin at the centre O, z along the upward vertical, x due south and y due east (Fig. 11). Then, with the notation of §2.2,

$$\dot{\xi} = \dot{x} \sin \lambda + \dot{z} \cos \lambda, \quad \dot{\eta} = \dot{y}, \quad \dot{\zeta} = -\dot{x} \cos \lambda + \dot{z} \sin \lambda,$$

when η and y are parallel. In place of $c\mathbf{n}$ in §2.3 we can write $-\omega^2(x, y, z)$ say, where ω is real and possibly time-dependent. By resolving on the new axes the equal vectors formed by the two sides of (3) in §2.3:

$$\left.\begin{aligned} \ddot{x} - 2\Omega\dot{y} \sin \lambda &= -\omega^2 x, \\ \ddot{y} + 2\Omega(\dot{x} \sin \lambda + \dot{z} \cos \lambda) &= -\omega^2 y, \\ \ddot{z} - 2\Omega\dot{y} \cos \lambda &= -\omega^2 z - g, \end{aligned}\right\} \tag{1}$$

with

$$x^2 + y^2 + z^2 = l^2.$$

These are linearized by treating x/l and y/l as first-order infinitesimals, and g and λ as constants. Then $\dot{z} \simeq (x\dot{x} + y\dot{y})/l$ and can be omitted from the second equation; from the third, ω^2 is constant to zeroth order and equal to g/l. This accuracy suffices for substitution in the other equations, which can afterwards be combined as

$$\left.\begin{aligned} \ddot{Z} + 2i\Omega\dot{Z} \sin \lambda + \omega^2 Z = 0 \\ \text{where} \quad Z = x + iy \quad \text{and} \quad \omega^2 = g/l. \end{aligned}\right\} \tag{2}$$

The Coriolis effect is represented by the term in $\dot{Z}$.

The relevant solution of this standard linear equation is an arbitrary real combination of particular integrals of exponential type:

$$\left.\begin{aligned} Z = (Ae^{i\nu t} + Be^{-i\nu t})e^{-i\Omega t \sin \lambda} \\ \text{where} \quad \nu^2 = \omega^2 + \Omega^2 \sin^2 \lambda. \end{aligned}\right\} \tag{3}$$

The complex constants A and B can be fitted to arbitrary initial values of position and velocity. To the present order of approximation $\nu \simeq \omega$ since $\Omega \ll \omega$. The projected path on the horizontal plane is a superposition of two uniform circular motions of possibly different amplitudes and phases, and with angular frequencies $-\Omega \sin \lambda \pm \nu$. The path is therefore an epicycle, certainly bounded, but closed only if these frequencies are commensurable (i.e. have a rational quotient). The form of (3) prompts the introduction of new coordinates $Z' = Z \exp (i\Omega t \sin \lambda)$, relative to which the path is plainly a fixed ellipse described in period $2\pi/\nu$. And, in conformity with this, one may verify the expected equation $\ddot{Z}' + \nu^2 Z' = 0$ representing elliptic harmonic motion; i.e., a combination of

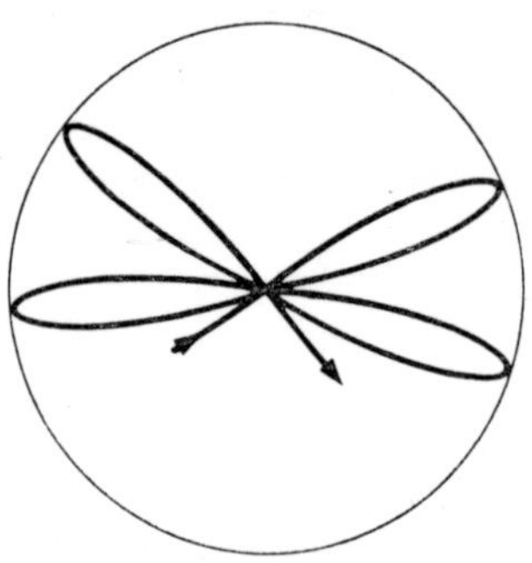

FIG. 12

perpendicular sinusoidal vibrations with equal frequencies but possibly different amplitudes and phases (the Z' axes are principal for the ellipse only when the phases differ by a quarter-period). The new axes turn round the upward vertical at rate $\Omega \sin \lambda$ left-handedly.

The predicted locus is well verified by experiments with pendulums *in vacuo*, or of such dimensions that air resistance is negligible (due originally to Foucault in 1851). When the pendulum is started so that it passes through the vertical, the ellipse degenerates to a line segment

$$Z' = ae^{i\alpha} \sin \nu t$$

where a and α are the arbitrary amplitude and orientation. Except on the equator the corresponding path relative to the earth is a succession of identical narrow loops (Fig. 12) with polar coordinates

$$r = a \sin \nu t, \quad \theta = \alpha - \Omega t \sin \lambda.$$

The angle between neighbouring loops, corresponding to one complete oscillation in time $2\pi/\nu$, $\simeq 2\pi\Omega \sin \lambda/\omega$ which is at most of the order of one minute of arc for practicable pendulums. At the poles the point of support is unaffected by the earth's spin, and the plane of swing has arbitrary constant orientation in a sidereal frame. This does not happen in other latitudes,† except trivially on the equator when the pendulum is set swinging in the equatorial plane.

What can be concluded from this and similar experiments? In the first instance they are a means of identifying certain types of constraint (with specifiable physical properties) as frictionless in the sense that the motion conforms with (2) in §2.3. Secondly, because of their duration, such experiments are well suited to measure $\mathbf{g}$ and hence $\mathbf{f}$ (by contrast with free fall). Next, and more basically, they support the proposition that an unconstrained particle, whatever its mass or velocity, always has this acceleration $\mathbf{f}$ with respect to the earth's centre, but relative to a frame in which the earth has a certain axial spin Ω. Moreover, the value of this Ω, as determined by the experiments, agrees with the value obtained astronomically for the earth's spin in a *sidereal* or *Newtonian* frame. And thus, in a last analysis, the experiments are concordant with the 'working approximation' adjoined to the gravitational postulate (§1.5): relative to a Newtonian frame any free body in the solar system has an acceleration *entirely* accountable for by the presence of other matter, and in particular this acceleration does not involve superficial terms of centripetal or Coriolis type which are removable by an appropriate change of frame.

2.5 Axisymmetric Constraint: Stability and Range of Motion

A frictionless constraint meriting detailed analysis, especially to illustrate methods of procedure, is a surface of revolution with axis vertical and fixed to the earth. It may for example be an actual container, or a simple pendulum whose suspension has negligible mass and either is perfectly stiff or is flexible and inextensible. Coriolis effects of the earth's rotation are disregarded, but the centripetal effect is retained implicitly in the gravity vector $\mathbf{g}$. Without Coriolis effects $\mathbf{g}$ is the acceleration relative to the earth that the particle would have if free, no matter what its speed (§2.2).

† One need only reflect that even the direction of the vertical generates a cone in a sidereal frame, and so *a fortiori* the string direction could not remain in one plane through O.

Accordingly its constrained acceleration relative to the earth is of form

$$\dot{\mathbf{v}} = \mathbf{g} + c\mathbf{n}$$

where $\mathbf{v}$ is the relative velocity (§2.3). Now the normal component of acceleration $\mathbf{n}\dot{\mathbf{v}}$ is equal to $\kappa\mathbf{v}^2$, where κ is the curvature of the normal section through the path (§2.3). Whence, by eliminating c,

$$\left.\begin{aligned}\dot{\mathbf{v}} &= (\kappa\mathbf{v}^2 - \mathbf{ng})\mathbf{n} + \mathbf{g} \\ &= \kappa v^2\mathbf{n} - \mathbf{n} \wedge (\mathbf{n} \wedge \mathbf{g}).\end{aligned}\right\} \qquad (1)$$

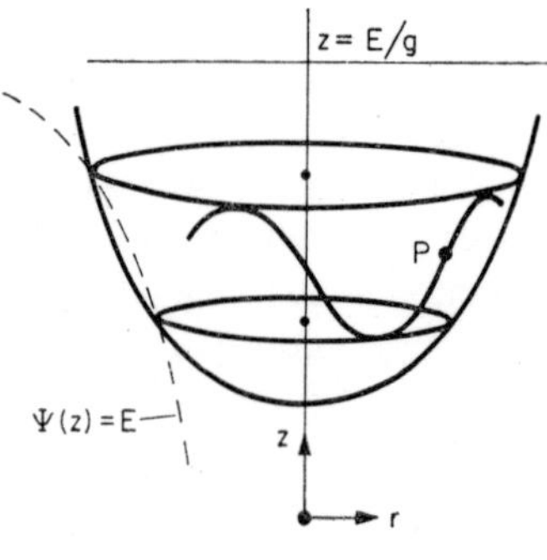

FIG. 13

The last term is just the gravity component tangential to the profile, while $\mathbf{n}$ here is in a meridian plane (Fig. 13). When the constraint is one-sided and unable to induce an additional acceleration towards itself, $c \geqslant 0$ if $\mathbf{n}$ is taken towards the side where the particle is, so that from the first equation (1)

$$\kappa v^2 \geqslant \mathbf{ng}. \qquad (2)$$

When this requirement ceases to be met, the particle leaves the constraint and moves freely under gravity.

The position P of the particle is specified by cylindrical coordinates (r, z) with z upwards along the axis of symmetry from any convenient origin; the profile $r(z)$ is assumed to have no corners but may have discontinuities in curvature. The circumferential component of velocity is $r\omega$ where ω is the rate of turn of the meridian plane through P at any moment, while the remaining component in the plane (tangential to the profile) is denoted by u. Since the circumferential component of acceleration vanishes,

$$\frac{1}{r}\frac{d}{dt}(r^2\omega) = 0 \quad \text{or} \quad r^2\omega = \text{constant}, \quad h \text{ say.} \tag{3}$$

This is a first integral of the equation of motion (1), with the interpretation that the moment of momentum has a constant component on (or about) the axis of symmetry. By forming the scalar product of $\mathbf{v}$ with (1), as in §2.3, another first integral is found:

$$\tfrac{1}{2}(u^2 + r^2\omega^2) + gz = \text{constant}, \quad E \text{ say,} \tag{4}$$

with the interpretation that the sum of the kinetic energy per unit mass and the potential of the local gravity field does not vary. Initial conditions of projection along the surface serve to fix E and h. When $h = 0$ the particle is confined to one meridian plane and oscillates finitely about the vertical. When $h \neq 0$ the particle cannot even approach the axis asymptotically, since the unbounded value of ω that would be implied by (3) is not permitted by (4) (except for merely academic surfaces such that $z \to -\infty$ with $r \to 0$). Equations (3) and (4) then determine ω and u^2 as functions of position only, and thence the orbit by a further integration.

By elimination of ω between (3) and (4):

$$\tfrac{1}{2}u^2 + \Psi(z) = E \quad \text{where } \Psi(z) = gz + \frac{h^2}{2r^2} \tag{5}$$

is a function of z for a given profile $r(z)$, with similar analytic properties. This equation governs the component motion in a meridian plane turning at the varying rate $\omega = h/r^2$ according to the position of the particle. Since $u^2 \geqslant 0$ the particle certainly remains in the range of z such that

$$\Psi(z) \leqslant E. \tag{6}$$

For given E and h this ordinarily yields a single compact annulus on the surface, when the cubic curve $\Psi(z) = E$ has just two intersections with the profile (Fig. 13); but with certain surfaces (having convolutions, for instance) there can be several annuli. Of course, the particle remains in the annulus that contains its initial position since it cannot leave without violating (6). Moreover, subject always to (2) when the constraint is one-sided, the particle travels monotonically across the entire annulus from one boundary to the other, since by (5) u cannot vanish when $\Psi(z) < E$. Its orbit contains no cusps but is tangential to both boundaries since ω does not vanish when $h \neq 0$; thus, when convenient, an orbit can be regarded as initiated by horizontal projection at a boundary (unless, exceptionally, this is approached asymptotically). Each segment of path between successive upper and lower points of tangency is identical, apart from reflection, since u^2 and ω are functions of height only (Fig. 13). Generally, the orbit does not close and would eventually 'fill' the permitted annulus.

Sufficiently near the upper boundary u^2 necessarily decreases with increasing z and so $d(u^2)/dz < 0$ there, while near the lower boundary $d(u^2)/dz > 0$. Whence, by differentiating (5),

$$h^2 \frac{dr}{dz} \gtrless gr^3 \tag{7}$$

where the upper (lower) inequality holds on the lower (upper) boundary. Conversely, this constitutes a criterion by which may be judged whether a particle begins to rise or fall when projected horizontally, say at height z_0 with moment of momentum h_0. Corresponding to this initial data we write

$$\Psi_0(z) = gz + \frac{h_0^2}{2r^2}, \qquad E_0 = \Psi_0(z_0). \tag{8}$$

In particular, when h_0 is such that the equality holds in (7), or equivalently

$$\frac{d}{dz}\Psi_0(z) = 0 \quad \text{when } z = z_0, \tag{9}$$

the particle necessarily continues at this level in circular orbit, and the annulus degenerates. This condition may also be derived by comparing tangential components on both sides of (1), the acceleration being purely centripetal and of amount $r\omega^2 = h_0^2/r^3$.

Suppose that the stationary value E_0 of $\Psi_0(z)$ at z_0 is a local minimum, so that $\Psi_0(z) - E_0$ is *positive-definite* near z_0; this implies a restriction on the profile curvature. Then (Fig. 14a) for every E sufficiently near to and greater than E_0, there exists an annulus $A(E)$ containing z_0 and such that

$$E_0 < \Psi_0(z) \leqslant E \quad \text{for} \quad z \neq z_0 \text{ in } A(E).$$

Moreover, and this is crucial in what follows, $A(E) \to 0$ as $E \to E_0$. Now consider a (non-circular) orbit resulting from any transitory disturbance of the steady motion (in position or velocity or both) leaving the moment of momentum unaltered. Provided such a disturbance is sufficiently small, the initial position falls within the annulus $A(E)$ that corresponds to the orbital E-value. By the remarks following (6), the orbit thereafter remains within $A(E)$, and this annulus $\to 0$ when $E \to E_0$ as the initial deviations in position and velocity are reduced indefinitely. This property in respect of arbitrary, vanishingly-small, disturbances is termed **orbital stability**;

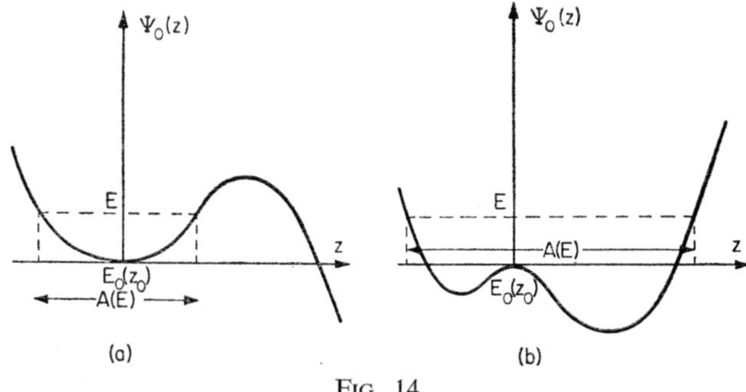

FIG. 14

the concept is by no means restricted to the present context. When, on the other hand, $\Psi_0(z) - E_0$ is not positive-definite (Fig. 14b) there must exist some compact annulus A containing z_0 (perhaps as one boundary) and such that

$$\Psi_0(z) \leqslant E_0 \text{ for } z \text{ in } A.$$

Then, necessarily, any orbit originating in A with $E > E_0$ and $h = h_0$ must have a range of at least A, since *a fortiori* $\Psi_0(z) < E$ in A. More especially, and this is the essential point, certain *vanishingly-small* disturbances can clearly be found to generate initial conditions of just this kind. Since A is *fixed*, such a response is mutually exclusive with stable behaviour, and is accordingly termed **orbital instability**.

We have thereby established that locally positive-definite $\Psi_0(z) - E_0$ is both necessary and sufficient for stability under disturbances leaving the moment of momentum unaltered. Under entirely general disturbances it is plainly necessary and sufficient that $\Psi_1(z) - E_1$ is locally positive-definite in z for each h_1 in an interval containing h_0. For, clearly, orbits which remain arbitrarily close to the steady-state level z_1 corresponding to h_1 are also arbitrarily close to z_0 as $h_1 \to h_0$. Notice, however, that the difference in mean circumferential speeds for any $h_1 \neq h_0$ eventually produces unbounded deviation in *azimuth* between contemporaneous positions in the circles z_1 and z_0.

Whether a function of one variable is locally positive-definite or not is recognizable from the first non-vanishing term in its Taylor expansion at the considered point: it will be if, and only if, the leading term is of even order and has a positive coefficient. With regard to $\Psi_0(z) - E_0$ let us suppose that this term is the second, as usually happens, and analyse a disturbed orbit (E, h_0) in more detail. We write the height of the particle as $z_0 + \zeta(t)$ where z_0 is the level of the steady motion with the same moment of momentum h_0, and retain only leading powers in the expansion of (5). With

$$u^2 = \{1 + (dr/dz)^2\}\dot{z}^2$$

and use of (8) and (9), there remains

$$\left\{1 + \left(\frac{gr_0^3}{h_0^2}\right)^2\right\}\dot{\zeta}^2 + \left(\frac{d^2\Psi_0}{dz^2}\right)_0 \zeta^2 = 2(E - E_0).$$

When there is stability the second derivative here is positive, according to the preceding general theorem and initial supposition. Evidently the expression is the first integral of a simple harmonic equation (whose solution can be written down merely by inspection of the integral itself):

$$\zeta(t) = \zeta(0) \cos (pt) + \frac{1}{p}\, \dot{\zeta}(0) \sin (pt) \tag{10}$$

with angular frequency p such that

$$p^2 = \left(\frac{d^2\Psi_0}{dz^2}\right)_0 \Big/ \left\{1 + \left(\frac{gr_0^3}{h_0^2}\right)^2\right\} \tag{11}$$

and an orbital E-value given by

$$E - E_0 = \frac{1}{2}\left(\frac{d^2\Psi_0}{dz^2}\right)_0 \left\{\zeta^2(0) + \frac{1}{p^2}\, \dot{\zeta}^2(0)\right\}.$$

When, on the other hand, there is instability the second derivative is negative and the solution is non-oscillatory:

$$\zeta(t) = \zeta(0) \cosh (nt) + \frac{1}{n} \dot{\zeta}(0) \sinh (nt) \tag{12}$$

where $n = ip$ is real. Generally this is unbounded as $t \to \infty$ and is therefore valid as an approximation only for comparatively small t. However, when $E = E_0$,

$$\frac{\zeta(t)}{\zeta(0)} = \exp \left\{ \frac{\dot{\zeta}(0) t}{\zeta(0)} \right\}$$

and either this is unbounded as $t \to \infty$, or the particle approaches the level z_0 asymptotically if it is initially tending in that direction. This illustrates the point that instability does not necessarily involve a finite response to all vanishingly-small disturbances, but only to certain kinds. The entire analysis is applicable to orbits (E, h_1), having different moments of momentum, merely by changing all subscripts from 0 to 1 and interpreting ζ as the deviation from the steady-state level z_1 appropriate to h_1. This level is defined by $d\Psi_1/dz = 0$, from which

$$z_1 - z_0 = 2(h_1 - h_0)g/h_0 \left(\frac{d^2\Psi_0}{dz^2} \right)_0$$

to first order.

Finally, by way of illustration, all the preceding results are specialized for a spherical pendulum. That there is but one permitted annulus is clear from the qualitative form of the curve $\Psi(z) = E$, which has no inflexion and is asymptotic to the downward axis and to the horizontal $z = E/g$ (Fig. 13). Let the radii to the boundaries from the point of support make angles α and β with the downward vertical $(0 < \alpha \leqslant \beta < \pi)$, and let l be the length of the pendulum. Then, by solving the relations

$$\Psi(-l \cos \alpha) = E = \Psi(-l \cos \beta)$$

for E and h, one obtains

$$\left.\begin{aligned} \frac{h^2}{2gl^3} &= \frac{\sin^2 \alpha \sin^2 \beta}{(\cos \alpha + \cos \beta)}, \qquad (\alpha + \beta < \pi) \\ \frac{E}{gl} &= \frac{(1 + \cos \alpha \cos \beta)}{(\cos \alpha + \cos \beta)} - (\cos \alpha + \cos \beta). \end{aligned}\right\} \tag{13}$$

When the string is flexible, (2) is the condition that it should remain taut. With $\kappa = 1/l$ and $\frac{1}{2}v^2 = E - gz$, there follows

$$z \leqslant 2E/3g$$

where z is measured upwards from the point of support. This is satisfied at the upper boundary, and hence elsewhere, when

$$(\cos \alpha + 2 \cos \beta)^2 \geqslant 1 - 9 \sin^2 \alpha$$

and so automatically when $\sin \alpha \geqslant 1/3$ for all speeds of projection. Under the moment of momentum h_0 the angular height γ of the possible steady state (conical pendulum) is obtainable from (9), or directly from (13) by putting $\alpha = \beta = \gamma$:

$$\frac{h_0^2}{gl^3} = \frac{\sin^4 \gamma}{\cos \gamma} \qquad (\gamma < \tfrac{1}{2}\pi).$$

This state is automatically stable (as it is whenever a profile is locally convex outwards) and the corresponding angular frequency is given by (11):

$$p^2 = \frac{g}{l} \cdot \frac{(1 + 3 \cos^2 \gamma)}{\cos \gamma}. \tag{14}$$

The meridian plane through the bob has spin $h_0/l^2 \sin^2 \gamma$, by (3), and therefore turns through an angle

$$\phi = 2\pi/\sqrt{(1 + 3 \cos^2 \gamma)} \tag{15}$$

in the period $2\pi/p$ of oscillation in a slightly disturbed motion. Unless γ happens to be such that $\phi/2\pi$ is a rational fraction between $\frac{1}{2}$ and 1 the orbit does not close. When the pendulum remains almost vertical,

$$\frac{\phi}{\pi} \simeq 1 + \frac{3}{8} \sin^2 \gamma \quad \text{as} \quad \gamma \to 0,$$

and so in two oscillations the bob turns through one complete revolution plus the second-order angle $\frac{3}{4}\pi \sin^2 \gamma$. A similar secular advance occurs in Foucault's experiment, superimposed on the Coriolis rotation; it is neglected in the first-order theory of §2.4.

2.6 Inertial Mass

It has been seen that an isolated particle moving in a field of acceleration, but partially under inexorable constraint, can still be treated within the framework of gravitational theory alone. This is no longer the case when the action of the constraint is conditional on the mass and motion of the particle, as for instance when the constraint is itself a free mass. For such phenomena the wider framework of Newtonian mechanics is required. Its full content will appear gradually; to begin with, we obtain a preliminary characterization of its basic elements through certain explanatory 'ideal' experiments.

In these the constraint is a perfect spring, by definition such that any connected particle P in continuous motion receives an *additional* acceleration whose magnitude α_P depends only on P and the current fractional extension ε (positive or negative) at that end. This additional acceleration is automatically the same in all frames of reference (by the invariance theorem in §2.1). It is observed to be independent of position, velocity, or local gravitational field, and its direction is always longitudinal when the spring is rectilinear and tangential when the spring is curved (inward when $\varepsilon > 0$ and outward when $\varepsilon < 0$). The $\alpha_P(\varepsilon)$ dependence is single-valued but not necessarily linear. Such behaviour can be observed, for example in rotary or oscillatory motion, with either 'hard' or 'soft' springs of tempered metal with negligible mass, at rates of strain such that the deformation is always effectively uniform along the length. When different particles P, Q, R . . . are attached in turn a remarkable result emerges (its impact inevitably blunted by long familiarity): the ratios $\alpha_P : \alpha_Q : \alpha_R$ . . . evaluated at the same terminal extension, *whatever this might be*, and irrespective of the motion, etc., are *invariant*. Moreover, these same ratios are obtained no matter how the material and construction of the spring are varied, subject to the stated requirements.

It can be concluded, therefore, that the ratios are intrinsic to the particles. The stipulation of this invariance for all such experiments, actual or potential, and for any motion or background field, constitutes a new theoretical postulate (cf. the concluding remarks in §1.5). Now let an arbitrary number, m_Q say, be assigned to a chosen particle Q. Then, for another particle P, the number $m_P = m_Q\alpha_Q(\varepsilon)/\alpha_P(\varepsilon)$ derived from any such experiment is called the **inertial mass** of P on the scale Q. The terminology is apt: the greater the inertia the slower is the change in velocity associated with a given spring configuration. The property is additive, in that the inertial mass of several particles combined is equal to the sum of their masses separately; this can be observed in further experiments where particles are attached together on one spring.

The international standard (1901) is a platinum–iridium cylinder known as the **prototype kilogramme** (a British pound is equal to 0·45359237 kg by recent definition). All inertial masses are ultimately compared with this standard, via certified copies and sub-multiples, and can be stated as a certain number of kilogrammes or of grammes (defined as 10^{-3} kg). This incidentally establishes another physical 'dimension' independent of length and time.

Any perfect spring can be calibrated absolutely by recording its $\alpha(\varepsilon)$ relation corresponding to a definite mass, say 1 kg. The inertial mass m_P of any other body P can then be determined by using the calibrated spring as a dynamometer, or more conveniently as a balance so as to maintain P at rest in the observer's frame of reference. The additional acceleration induced by the spring is then necessarily equal and opposite to the local initial acceleration, g say, in free fall relative to that frame; in the terrestrial frame, for example, g is local gravity. Thus $m_P = \alpha(\varepsilon)/g$ kg where ε is the steady extension or compression and depends on g. Generally, the product $m_P g$ is called the **weight** of P, with dimensions of [mass][length][time]$^{-2}$. When, in particular, the frame of reference is such that the corresponding g vanishes, the body can be said to be 'weightless'. While the weights of all bodies depend on both position and reference frame, their ratios under the same conditions do not, being equal to the respective mass ratios.

A dynamometer, with known relation $\alpha(\varepsilon)$, is conversely capable of measuring the additional acceleration of a known mass, m kg, connected to it. In particular, when used as a balance in a gravimetric survey, it measures local gravity as $g = \alpha(\varepsilon)/m$. Or, again, it can be used as a primitive accelerometer in an inertial guidance system, designed to locate position $\mathbf{r}(t)$ in a Newtonian frame. The gravitational field $\mathbf{f}(\mathbf{r})$ is known in advance; the magnitude $\alpha(t)$ is furnished as a function of time by the recorded extension $\varepsilon(t)$; the spring direction in the Newtonian frame is inferred from the orientation of the vehicle in that frame (determined directly either by a gyroscopic device or indirectly by measuring the relative inclination of the field $\mathbf{f}$). The required position then follows by continuous integration of $\ddot{\mathbf{r}} = \mathbf{f}(\mathbf{r}) + \boldsymbol{\alpha}(t)$ starting from given initial data.

We return to the invariance postulate and supporting experiments. Since only the terminal fractional extension of a perfect spring is relevant here, conditions at the end remote from the considered body, P say, may be varied at will. Suppose, for example, that a second body Q is attached there. Its motion also is subject to the postulate, and accordingly $\alpha_Q/\alpha_P = m_P/m_Q$ when the terminal extensions are equal, in particular when the spring is uniformly strained. This last is, indeed, characteristic for a perfect spring either when it is rectilinear and free or smoothly constrained and curved. Thus, if passed round a fixed frictionless support so that the ends hang vertically, the actual acceleration a of the system from Q towards P is such that $g + a = \alpha_Q$ and $g - a = \alpha_P$, or $(g + a)/(g - a) = m_P/m_Q$, or $a/g = (m_P - m_Q)/(m_P + m_Q)$; this, of course, is the principle of Atwood's machine, which affords a simple but restricted test of the invariance property. When, on the other hand, the spring remains rectilinear, we can express both this and the ratio of magnitudes in a single vector equation

$$m_P\boldsymbol{\alpha}_P + m_Q\boldsymbol{\alpha}_Q = \mathbf{0} \tag{1}$$

remembering that the senses of the two additional accelerations are opposed.† Shortly stated, the equation asserts that the acceleration of the inertial mass-centre is unaffected by spring constraint; the Fletcher trolley experiment illustrates this general principle. For the time being, the analogy with (5) in §1.13 for a free pair of gravitating bodies remains formal.

The preferred way of measuring inertial mass is by 'weighing' on a beam balance. The validity of the method stems from a further

† This property of connected pairs of particles can serve as an alternative characterization and definition of inertial mass. However, it is neither so attractive nor so expedient as that in the text, since the prototype situation is less general and has an air of artificiality. Nevertheless, it has been adopted by many authorities, among them Love, Whittaker, and Milne. It was first codified by Mach (1868): "*Experimental proposition:* Bodies set opposite each other induce in each other, under certain circumstances to be specified by experimental physics, contrary accelerations in the direction of their line of junction. *Definition:* The mass-ratio of any two bodies is the negative inverse ratio of the mutually induced accelerations of those bodies. *Experimental proposition:* The mass-ratio is independent of . . .", etc. A partial development of this approach, and a critique of Newton's methodology, is set out in Mach's *The Science of Mechanics* (1883), especially Chapter II, Sections (v), (vi) and (vii) (The Open Court Publishing Company 1960, with a new introduction by K. Menger to the revised English translation).

experimental fact: the equilibrium of the balance *in vacuo* is undisturbed when bodies P and Q of equal weight (but otherwise arbitrarily dissimilar) are placed in the scale-pans. Such continuing equilibrium can subsist in any frame of reference where the direction of the corresponding *initial* acceleration in free fall is sensibly uniform over the region occupied by the balance. If g_P and g_Q are the known magnitudes of this acceleration at the scale-pans then $m_P/m_Q = g_Q/g_P$, from which either mass is determined when the other is given. Ordinarily, of course, the scale-pans are at the same level and the frame is terrestrial, so that $g_P = g_Q =$ local gravity and $m_P = m_Q$.† However, balances of large dimensions with scale-pans at different levels are occasionally used to demonstrate, conversely, the variation of gravity with height, both masses being known.

Suppose, now, that the field of acceleration at one scale-pan of an ordinary balance is augmented by some unspecified amount f by placing directly beneath the pan a uniform sphere of large mass m at centroidal distance r (Poynting's experiment). Then, if two standard kilogrammes (say) were previously balanced, their equilibrium could only be maintained by adding to the other pan a small mass f/g kg. Whence, g being known, the actual value of f is determined; and by varying r the inverse-square dependence itself can be investigated. This done, the gravitational mass of m in metric units is obtained as $\mu = r^2 f$ and finally the ratio μ/m. Alternatively, suppose a standard kilogramme (say) is maintained without horizontal acceleration in the field of the sphere m by a horizontal calibrated spring; then, with the previous notation, μ/m is found as $r^2\alpha(\varepsilon)/m$. Equivalently, but with greater sensitivity, a torsion balance can be used (Cavendish–Boys' experiment).

By such methods μ/m is found to be constant, the same for all bodies, within experimental error. This striking fact, linking

† Inertial mass could be *defined* operationally as the number of standard units needed to balance a given body at the same height. But its dynamical significance would still rest on the observed invariance of the acceleration ratios under spring constraint; their equality with the mass-ratios would be a fact of *observation*, not a definition as before.

The apparent circularity of Newton's definition of mass as volume × density has often been remarked. However, it seems likely that 'density' then had the meaning of the modern 'specific gravity' (*Principia:* Cajori's Appendix, Note 11). If this be so, Newton's definition is saved, since it essentially involves balancing unit volume of body against a number of unit volumes of water.

apparently distinct sets of phenomena underlying the respective definitions of μ and m, is the keystone of Einstein's general theory of relativity. In classical mechanics the invariance of this relation is postulated universally, even for celestial bodies whose inertial mass-density cannot at present be determined independently. We write

$$\mu = \gamma m. \tag{2}$$

where γ is known as the **gravitational constant**, with a currently accepted value of $6{\cdot}670 \times 10^{-8}\ \text{cm}^3/\text{g sec}^2$. It follows that the additional acceleration induced by ordinary bodies is completely negligible compared with terrestrial gravity; for example, at the surface of a uniform sphere of radius r and inertial mass-density ρ, the field $f = \frac{4}{3}\pi\gamma\rho r$ is only $2{\cdot}794 \times 10^{-4}\ \text{cm/sec}^2$ for the typical values $r = 100$ cm, $\rho = 10\ \text{g/cm}^3$ (about the density of silver).

From the known gravitational mass of the earth (§1.18) its inertial mass is derived as $5{\cdot}976 \times 10^{27}$ g, to the number of significant figures in γ. In terms of a nominal volume given by the reference spheroid (§1.18), namely $\frac{4}{3}\pi a^2 c = 1{\cdot}0833\ldots \times 10^{27}\ \text{cm}^3$, the mean inertial density is $5{\cdot}516\ \text{g/cm}^3$. The earth's crust is found to have a density of about $2{\cdot}7\ \text{g/cm}^3$ only, however. It is generally accepted that the density increases inwards monotonically, first gradually through the mantle up to about $6\ \text{g/cm}^3$ at a depth of nearly 3000 km, and then more rapidly through the core with uncertain variation about a mean of some 12 or $13\ \text{g/cm}^3$. The possible distributions of density are fairly closely circumscribed by seismological and astronomical data through theories of the internal and external fields.

2.7 Force: Second and Third Laws of Motion

The outcome of experiments on ideal spring constraint, of whatever kind, may be re-phrased in the following way. The additional acceleration $\boldsymbol{\alpha}_P^S(\varepsilon)$ induced in any (small) body P by any spring S with any local strain ε is decomposable as $\mathbf{F}^S(\varepsilon)/m_P$, where m_P is an invariant scalar for that body and $\mathbf{F}^S(\varepsilon)$ is an invariant vector function of extension for that spring. m_P has been called the inertial mass of body P (on some scale), but no term for $\mathbf{F}^S(\varepsilon)$ has been introduced. It is clearly convenient to have one in order to emphasize the decomposition property. Accordingly, $\mathbf{F}^S(\varepsilon)$ is called the **force** due to (terminal) strain ε in spring S; its direction is always tangential to the end, while the dependence of its magnitude on ε can be recorded once and for all by calibration.

One can speak of applying a *given* force to any chosen body or to several bodies in turn: for the amount of force is *independently recognizable* from the configuration (spring + extension), without reference to the particular acceleration induced. All this being understood, the relationship can be expressed briefly, and for any frame of reference, as

inertial mass × additional acceleration = acting force, (1)

for any (small) body and any spring. This is a modern paraphrase of Newton's **Second Law of Motion**; its status is clearly empirical since it relates three terms which have been separately defined *a priori.*† Exceptionally, in regard to the standard body (used both for weighing and calibrating), the second law simply acts as a prescription of an absolute scale of force. The c.g.s. unit on this scale is a **dyne**, defined to correspond to an acceleration of 1 cm/sec^2 in a mass of 1 g. The m.k.s. unit of 10^5 dynes $\equiv$ 1 newton is also widely used, and corresponds to an acceleration of 1 metre/sec^2 in a mass of 1 kg.

If, now, the acceleration $\mathbf{f}$ of some gravitational field is included on the left side of (1), the product $m\mathbf{f}$ must similarly be added vectorially on the right. By a natural extension of terminology this product is spoken of as the 'force due to the field', and then the right side of (1) refers to the resultant of the spring and field forces. This usage must nevertheless not blur the fact that gravitational force depends on the actual mass affected, and not only on the determining configuration of matter; for the latter fixes the acceleration and not the force (by contrast with spring action). To emphasize this dependence it is customary to speak of gravitational **body-force.**††

† "Change of motion is proportional to the motive force and is made in the direction of the right line in which that force is impressed." Mach's dismissal of the second law as a mere definition of force is unjustified, and arises from his pre-occupation with pairs of bodies and their mutual actions. The essential fact, that a force magnitude can be specified operationally without reference to its effects, is entirely missing from his analysis of the phenomena.

†† The physiological effects of spring force and body-force are also markedly different. A gravitational field produces no stress gradient within a finite mass when the induced acceleration is effectively uniform throughout its volume. By contrast, spring force applied on the surface does produce a gradient, and hence observable deformation, etc. Compare, for example, the sensations experienced by space travellers when in 'free orbit' and at 'take-off', even though the additional accelerations may be comparable or even equal.

The content of (1) can be further enlarged through another observation. Under several springs attached together to a particle, the additional acceleration is found to be the vector sum of the separate additional accelerations when the springs are attached singly and have the same respective extensions. By a simple theorem in kinematics this property can clearly be replaced by the *postulate* that an independent additional acceleration is induced by each spring, irrespective of the others or of the local field. This **principle of independent action** can equally well be phrased in terms of the spring forces acting on the particle: their combined effect is postulated as equivalent to that of a single force equal to their vector resultant.† The form of words in (1) still stands, therefore, with the right side interpreted as the resultant of all the separately given forces.

Next, if an additional acceleration $\mathbf{c}$ attributable to any other kind of constraint, active or passive, is added on the left of (1), the right side must be augmented by the product $m\mathbf{c}$. This product is called a **contact force**, not only for linguistic uniformity, but since the constraint could in principle be maintained by an actual spring (operating servo-mechanically, for instance) and, by the principle of independent action, the force registered would indeed be $m\mathbf{c}$. However, this force is not *given* but depends on the entire dynamical situation. In particular, when the constraint is active and preordained (§2.3), the term essentially plays only a kinematic role in the equation of motion of the considered particle.

Suppose, on the other hand, that the constraint is passive, namely direct contact with another particle whose own motion is also to be found. This is a new situation, calling for a new principle of procedure. It is supplied by Newton's **Third Law of Motion,**†† which might be re-stated as follows: the additional motions directly arising from the contact of two bodies are accountable for by contact forces which are oppositely directed in the same line and have equal magnitudes. This implies, for extended bodies (§2.9) as well as particles, that the two mass-centres receive parallel but contrary additional accelerations in inverse ratio to the masses. The motion

† Newton's attempt at an *a priori* demonstration of the parallelogram rule tacitly assumed this independence, which is indeed only given by experiment (*Principia:* first corollary to Law I).

†† "To every action there is always opposed an equal reaction".

of the combined mass-centre, therefore, is not directly affected by the contact.

It will be recalled that similar properties were derived for a pair of particles connected by a rectilinear spring (§2.6) and also in respect of their mutual fields (§1.13) [the inertial and gravitational mass-centres are indistinguishable, by (2) in §2.6]. The respective pairs of forces are each equal and opposite in the same line. Such similarities, and others in regard to electrostatic or magnetostatic fields, gave rise to the time-honoured formulation of the third law: action and reaction are equal and opposite. This, however, is a composite statement and not an irreducible and specific axiom in a theoretical framework. As the present development makes clear, the statement subsumes several distinct phenomena, some already covered by the second law and the gravitational hypotheses. For this reason the restricted version of the third law is preferred here.

2.8 Fictitious Body-forces

We are now in a position to formulate the equation of motion for a particle under co-existent forces of any kind. Relative to a Newtonian frame the entire, apparent, acceleration of a *free* particle is given by the gravitational field of the solar system (first law of motion, §1.6). Consequently, the apparent acceleration of a constrained particle is accounted for by the recognizable forces present (gravitational, spring, contact, non-mechanical). Thus, *in any Newtonian frame*, the general equation of motion can be put shortly as

$$\text{mass} \times \text{apparent acceleration} = \text{resultant of actual forces}. \quad (1)$$

This may, indeed, be adopted as an alternative statement of the second law, fully equivalent to (1) in §2.7 given the prior first law.

Seen from any other frame of reference, the entire acceleration is generally not equal to the resultant of the invariant accelerations associated with recognizable acting forces. For, as shown in §2.1, the apparent accelerations in two frames of reference differ by kinematic terms connected with the relative motions of the frames and the particle. Let $\mathbf{r}$ denote position in an adopted frame α whose spin is $\mathbf{\Omega}(t)$ against the background of a Newtonian frame β. Further, let the chosen origin O in α have acceleration $\boldsymbol{\alpha}_O(t)$ in β.

Then, from (i), (ii) and (iii) in §2.1 or generally from (8) in §3.1,

$$\text{apparent acceleration in } \beta - \text{apparent acceleration in } \alpha$$
$$= \mathbf{\Omega} \wedge (\mathbf{\Omega} \wedge \mathbf{r}) + \dot{\mathbf{\Omega}} \wedge \mathbf{r} + 2\mathbf{\Omega} \wedge \dot{\mathbf{r}} + \boldsymbol{\alpha}_O,$$

where $\dot{\mathbf{r}}$ denotes velocity relative to α and $\dot{\mathbf{\Omega}}$ denotes the spin rate of change (relative to either α or β). Whence, by re-arrangement of (1), the equation of motion in the adopted frame is

$$m\ddot{\mathbf{r}} = \mathbf{F} + m\mathbf{\Omega} \wedge (\mathbf{r} \wedge \mathbf{\Omega}) + m\mathbf{r} \wedge \dot{\mathbf{\Omega}} + 2m\dot{\mathbf{r}} \wedge \mathbf{\Omega} - m\boldsymbol{\alpha}_O \quad (2)$$

where $\mathbf{F}$ is the resultant of the recognizable, actual, forces.

The extra terms now appearing on the force side of the equation are proportional to the mass and are given functions of time, position and velocity. Accordingly they are called the **fictitious body-forces** associated with α. This terminology allows the equation of motion in any frame to be stated in a uniform way:

$$\text{mass} \times \text{apparent acceleration}$$
$$= \text{resultant of all forces (actual or fictitious).} \quad (3)$$

In order to pass from the equation of motion in α to that in some other frame α' we simply add to either side the quantity

$$\text{mass} \times (\text{apparent acceleration in } \alpha'$$
$$- \text{apparent acceleration in } \alpha).$$

When this contribution to the force side is evaluated explicitly, as in (2), the extra terms can be said to be the fictitious body-forces for α' *vis-à-vis* α. Clearly, these are equal to the difference of those for α' and α respectively *vis-à-vis* β, or any other frame.

In detail the terms in (2) may be described and named as follows:

(i) $m\mathbf{\Omega} \wedge (\mathbf{r} \wedge \mathbf{\Omega})$ is the **centrifugal body-force**, parallel to the instantaneous plane of rotation and radially outwards from the spin axis placed at O.

(ii) $m\mathbf{r} \wedge \dot{\mathbf{\Omega}}$ is the **Euler body-force**, perpendicular to the radius and in a plane transverse to the direction of spin-rate $\dot{\mathbf{\Omega}}$.

(iii) $2m\dot{\mathbf{r}} \wedge \mathbf{\Omega}$ is the **Coriolis body-force**, perpendicular to the component of apparent velocity on the plane of rotation.

(iv) $-m\boldsymbol{\alpha}_O$ is the **translational body-force**, opposite in direction to the acceleration of O in a Newtonian frame.

When the spin axis has constant orientation in a Newtonian frame, the forces (i), (ii) and (iii) are coplanar. Their senses and directions can then be indicated on the plane of rotation, in relation to the projection of the apparent path in α (Fig. 15).

Let the gravitational acceleration of the considered particle by the entire solar system be $\mathbf{f}$ relative to O and a sidereal frame. The corresponding gravitational force is $m(\boldsymbol{\alpha}_O + \mathbf{f})$. As judged from the

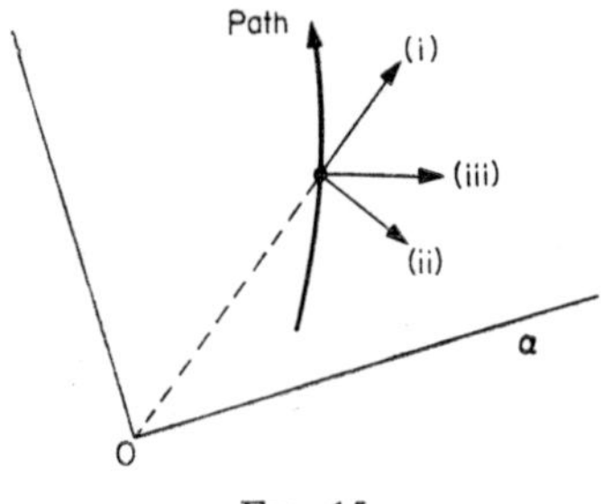

FIG. 15

adopted frame, the initial acceleration of the particle in free fall from $\mathbf{r}$ would be

$$\mathbf{g} = \mathbf{f} + \boldsymbol{\Omega} \wedge (\mathbf{r} \wedge \boldsymbol{\Omega}) + \mathbf{r} \wedge \dot{\boldsymbol{\Omega}} \tag{4}$$

by specializing (2), or directly from the prior kinematic connexion. The product $m\mathbf{g}$ has already been defined as the weight of the particle, in the generic position and with respect to the adopted frame (§2.6). Also, $\mathbf{F} - m(\boldsymbol{\alpha}_O + \mathbf{f})$ in (2) is the resultant of the actual non-gravitational forces. Therefore, having regard to (4), the general equation of motion can be stated as

mass × apparent acceleration

= weight + non-gravitational and Coriolis forces. (5)

In the event that the origin O is chosen to be the mass-centre of a freely-moving celestial body, natural or artificial, $\mathbf{f}$ is the field at P due to this body plus the differential acceleration of P with respect to O attributable to the rest of the solar system (including neighbouring matter not regarded as part of the chosen body). When O is the centre of a planet and P is near its surface, this differential acceleration makes a contribution to $\mathbf{g}$ which can usually be neglected (cf. §2.2). Otherwise, however, this cannot be taken for granted, for example when the frame is carried by a space vehicle in free flight.

2.9 Continuum Dynamics: Preliminary Notions

The principles so far stated do not relate to extended bodies, except in respect of the acceleration of the mass-centre in a gravitational field (§1.13). The Newtonian laws of motion themselves, either as expounded in the *Principia* or as re-interpreted here (§§2.6 and 2.7), have to do only with 'mass-points' whose geometry and rotational motion are by implication unimportant. A general basis for dynamics did not emerge for nearly a century after the *Principia*:† its final expression is due to Euler (1776).

Analogies with a system of mass-points nevertheless prove fruitful, as it happens, but an actual body is abstracted as a *continuum*. All state variables are conceptual constructs required to be at least piecewise-continuous functions of position; they are to be set in correspondence with experimental distributions in the way already explained for gravitational mass-density (§1.13). We begin by introducing some overall quantities similar to those encountered in the n-body problem (§1.10).

The **inertial mass-centre** of a continuum (Euler 1765) is that unique point, from now on denoted by G, with the property

$$\int \boldsymbol{\rho}\, dm = \mathbf{0} \tag{1}$$

where the integral extends over the entire mass and dm denotes an infinitesimal element in position $\boldsymbol{\rho}$ with respect to G. This is analogous to (2) in §1.13 and is indeed identical in view of the connexion (2) in §2.6. Once again, we warn that a mass-centre is not usually coincident with the same mass element at all times. But it clearly is when the whole body is either rigid or undergoes homogeneous deformation (i.e. the cartesian coordinates of all elements

† Newton did not discuss rotational motion, except in the remarkable but obscure treatment of the earth's precession in the luni-solar field. As for linear motion of extended bodies Newton was again exclusively concerned with gravitational action, and in particular implicitly *took for granted* that a mass-centre moves as a particle would under the force resultant (for instance in regard to the mutual orbit of two attracting spheres). For the gravitating system of mass-points he proved merely that the mass-centre is not additionally accelerated, *other forces being excluded* (Book I: Corollary IV).

at time t are the same linear forms in the initial coordinates, with coefficients depending only on t).

The **linear momentum** of a continuum is

$$\int \mathbf{v}dm = m\mathbf{v}_G \tag{2}$$

where $\mathbf{v}$ is the apparent velocity of the element dm in the chosen frame of reference, $\mathbf{v}_G$ is the velocity of G, and m is the total mass. This equality put as

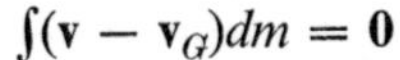

$$\int(\mathbf{v} - \mathbf{v}_G)dm = \mathbf{0}$$

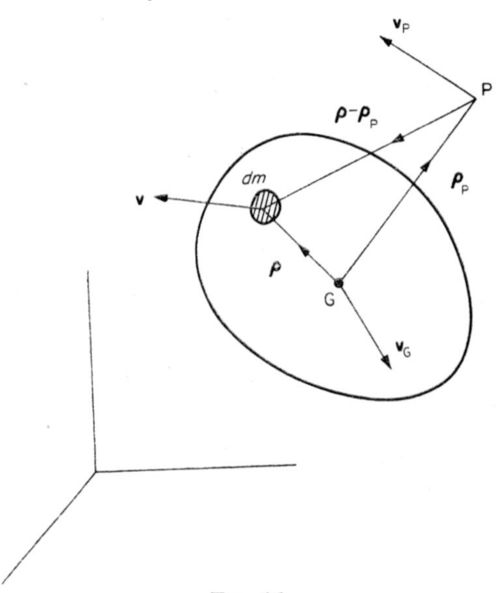

FIG. 16

can be seen as the time-derivative of (1), since $\dot{\boldsymbol{\rho}} = \mathbf{v} - \mathbf{v}_G$; it expresses the vanishing of the momentum of the motion relative to G. Similarly,

$$\int \dot{\mathbf{v}}dm = m\dot{\mathbf{v}}_G \tag{3}$$

is the linear mass-acceleration or rate of change of momentum.

The **moment of momentum** (often inappropriately called angular momentum), taken about a point P, is defined as

$$\mathbf{h}(P) = \int(\boldsymbol{\rho} - \boldsymbol{\rho}_P) \wedge \mathbf{v}dm \tag{4}$$

where $\boldsymbol{\rho}_P$ is the position of P with respect to G (Fig. 16); this concept also is essentially due to Euler. When P coincides with G, (4) becomes

$$\mathbf{h}(G) = \int \boldsymbol{\rho} \wedge \mathbf{v}\,dm = \int \boldsymbol{\rho} \wedge \dot{\boldsymbol{\rho}}\,dm = \mathbf{h}_{\text{rel}}(G), \text{ say,} \tag{5}$$

since the left side of (1) has a null product with any vector, here $\mathbf{v}_G$. Hence, in computing moment of momentum about the mass-centre, either absolute or relative velocity may be used. Knowing the linear momentum as well, we can find the moment of momentum about any other point (not necessarily in the body) directly as

$$\mathbf{h}(P) = \mathbf{h}(G) - \boldsymbol{\rho}_P \wedge m\mathbf{v}_G \tag{6}$$

in replacement of (4), since $\boldsymbol{\rho}_P$ can be taken outside the integral. The first right-hand term is sometimes called the rotational part of the moment of momentum (arising from the motion relative to G) and the second term the orbital part (as if for a mass-point at G). The moment about P of the motion *relative* to P is, by analogy with (6),

$$\mathbf{h}_{\text{rel}}(P) = \mathbf{h}(G) - \boldsymbol{\rho}_P \wedge m(\mathbf{v}_G - \mathbf{v}_P).$$

This is more informative as

$$\mathbf{h}(P) = \mathbf{h}_{\text{rel}}(P) - \boldsymbol{\rho}_P \wedge m\mathbf{v}_P, \tag{7}$$

which makes plain that absolute or relative velocity can be used indifferently when, and only when, P is currently at rest or at G or moving in the line GP.

For the vector field of mass-acceleration the analogy to (6) is

$$\int (\boldsymbol{\rho} - \boldsymbol{\rho}_P) \wedge \dot{\mathbf{v}}\,dm = \int \boldsymbol{\rho} \wedge \dot{\mathbf{v}}\,dm - \boldsymbol{\rho}_P \wedge m\dot{\mathbf{v}}_G, \tag{8}$$

with the use of (3). By (4) the left side here is equal to

$$\frac{d}{dt}\mathbf{h}(P) + \int (\mathbf{v}_P - \mathbf{v}) \wedge \mathbf{v}\,dm$$

since $d(\boldsymbol{\rho} - \boldsymbol{\rho}_P)/dt = \mathbf{v} - \mathbf{v}_P$, and so to

$$\frac{d}{dt}\mathbf{h}(P) + \mathbf{v}_P \wedge m\mathbf{v}_G \tag{9}$$

by (2), since $\mathbf{v}_P$ can be taken outside the integral. The moment of mass-acceleration therefore coincides with the rate of change of moment of momentum when and only when P or G is instantaneously at rest, or their paths are currently parallel or tangential.

More particularly, by taking moments about the mass-centre itself, there follows the far-reaching identity

$$\int \boldsymbol{\rho} \wedge \dot{\mathbf{v}} dm = \frac{d}{dt}\mathbf{h}(G) = \frac{d}{dt}\mathbf{h}_{\mathrm{rel}}(G). \tag{10}$$

This incidentally makes explicit the right side of (8) for general P:

$$\frac{d}{dt}\mathbf{h}_{\mathrm{rel}}(G) - \boldsymbol{\rho}_P \wedge m\dot{\mathbf{v}}_G. \tag{11}$$

Finally, the substitution of (7) in (9) gives yet another expression for the moment of mass-acceleration:

$$\frac{d}{dt}\mathbf{h}_{\mathrm{rel}}(P) - \boldsymbol{\rho}_P \wedge m\dot{\mathbf{v}}_P, \tag{12}$$

remembering once more that $\dot{\boldsymbol{\rho}}_P = \mathbf{v}_P - \mathbf{v}_G$.

Next, we specify the relevant forces. These are of two kinds: body-forces (gravitational, fictitious) distributed throughout the continuum, and applied forces (spring, contact) distributed over its surface S. Let the total body-force per unit mass at any interior point be $\boldsymbol{\gamma}$ (the **body-force density**), and let the applied force per unit area at each surface point be $\boldsymbol{\tau}$ (the **surface traction**). The corresponding force and couple resultants are

$$\left.\begin{aligned} \mathbf{R} &= \int \boldsymbol{\gamma} dm + \int \boldsymbol{\tau} dS, \\ \mathbf{M}(G) &= \int \boldsymbol{\rho} \wedge \boldsymbol{\gamma} dm + \int \boldsymbol{\rho} \wedge \boldsymbol{\tau} dS, \end{aligned}\right\} \tag{13}$$

$$\mathbf{M}(P) = \mathbf{M}(G) - \boldsymbol{\rho}_P \wedge \mathbf{R}. \tag{14}$$

It is to be understood that idealized point-loads and point-torques, typically $\mathbf{L}$ and $\mathbf{T}$ respectively, are explicitly included in the surface integrals. They are formally obtained by imagining the traction magnitude to be increased indefinitely over a vanishingly small area σ around a point π on S in such a way that

$$\int \boldsymbol{\tau} d\sigma \to \mathbf{L}, \quad \int (\boldsymbol{\rho} - \boldsymbol{\rho}_\pi) \wedge \boldsymbol{\tau} d\sigma \to \mathbf{T},$$

as $\sigma \to 0$. These together imply

$$\int \boldsymbol{\rho} \wedge \boldsymbol{\tau} d\sigma \to \mathbf{T} + \boldsymbol{\rho}_\pi \wedge \mathbf{L}.$$

When the adopted frame translates with a free celestial body and has a (possibly different) spin $\mathbf{\Omega}$ against the sidereal background,

$$\boldsymbol{\gamma} = \mathbf{g} + 2\mathbf{v} \wedge \mathbf{\Omega}$$

as shown in §2.8. Here $\mathbf{v}$ is the apparent velocity and $\mathbf{g}$ is the initial acceleration in free fall from rest, given by (4) there. For any moderate spin (such as the earth's), and any continuum of ordinary dimensions, it is fully adequate to treat $\mathbf{g}$ as uniform. Its total contributions to the resultants are then effectively those of a force $m\mathbf{g}$ regarded as located at G, no matter what the orientation of the continuum, since

$$\int \boldsymbol{\rho} \wedge \mathbf{g}\,dm \simeq \int \boldsymbol{\rho}\,dm \wedge \mathbf{g} = \mathbf{0}$$

by the property (1).† [Of course, this is far from saying that the imagined force is mechanically equivalent in all respects to the distributed weight.] The resultants can now be written, more explicitly, as

$$\left.\begin{aligned} \mathbf{R} &= m\mathbf{g} + \int \boldsymbol{\tau}\,dS + 2m\mathbf{v}_G \wedge \mathbf{\Omega}, \\ \mathbf{M}(G) &= \int \boldsymbol{\rho} \wedge \boldsymbol{\tau}\,dS + 2\int \boldsymbol{\rho} \wedge (\dot{\boldsymbol{\rho}} \wedge \mathbf{\Omega})\,dm, \end{aligned}\right\} \quad (15)$$

having regard to (1) and (2). In the earth's frame of reference, in particular, the Coriolis force and couple are normally both negligible [but see §§2.2, 2.4 and 3.9 (ii)].

Even when the considered continuum is only part of a naturally coherent body, the mechanical influence of the remainder is still conceived to be fully represented by a distributed traction over the interface. The local tractions of either part on the other are, of course, taken to be equal and opposite. This concept replaces, on the macroscopic level, what would be described in fine detail as field forces between atoms and electrons. In ultimate justification stands the whole of engineering science.

2.10 Continuum Dynamics: Equations of Motion

The linear and moment resultants over a continuum have now been defined for the respective vector fields of force and mass-acceleration. The entire classical mechanics of continua (both solid

† Only then is it permissible to speak of G as a 'centre of gravity'.

and fluid) stems from the postulate that the resultants in these two fields are always in balance:

$$\left.\begin{aligned} \int \dot{\mathbf{v}} dm &= \int \boldsymbol{\gamma} dm + \int \boldsymbol{\tau} dS, \\ \int \boldsymbol{\rho} \wedge \dot{\mathbf{v}} dm &= \int \boldsymbol{\rho} \wedge \boldsymbol{\gamma} dm + \int \boldsymbol{\rho} \wedge \boldsymbol{\tau} dS. \end{aligned}\right\} \tag{1}$$

It is enough to assert this in a single frame of reference. The analogue of (1) then follows for any other frame merely by adding to $\dot{\mathbf{v}}$ the appropriate difference in the respective apparent accelerations and to $\boldsymbol{\gamma}$ precisely the same quantity interpreted as a fictitious body-force density (§2.8). The integrals over the mass thus change by equal amounts in each equation, while the integrals over the surface are unaffected since traction is an invariant (§2.7). It is also enough to assert (1) for the moment about a single point, here the mass-centre, since the moments about any other are then automatically in balance:

$$\int (\boldsymbol{\rho} - \boldsymbol{\rho}_P) \wedge \dot{\mathbf{v}} dm = \int (\boldsymbol{\rho} - \boldsymbol{\rho}_P) \wedge \boldsymbol{\gamma} dm + \int (\boldsymbol{\rho} - \boldsymbol{\rho}_P) \wedge \boldsymbol{\tau} dS, \tag{2}$$

by multiplying the first equation by $-\boldsymbol{\rho}_P$ and adding to the second. Again, it suffices to replace the second of (1) by the assertion that the couple resultant is in balance with the rate of change of the moment of momentum about a point *fixed* in the chosen frame [(9) in §2.9, with $\mathbf{v}_P = \mathbf{0}$]. In this form, and in the context of a Newtonian frame (where the fictitious body-forces vanish), the postulate is due to Euler (1776).

Concise expressions of (1) are

$$m\dot{\mathbf{v}}_G = \mathbf{R}, \qquad \frac{d}{dt}\mathbf{h}_{\text{rel}}(G) = \mathbf{M}(G), \tag{3}$$

by (3) and (10) in §2.9. By direct combination,

$$\frac{d}{dt}\mathbf{h}_{\text{rel}}(G) - m\boldsymbol{\rho}_P \wedge \dot{\mathbf{v}}_G = \mathbf{M}(P) \tag{4}$$

which corresponds to (2), as noted in (11) of §2.9. According to the first of (3) the mass-centre G moves as if it were an isolated particle with all forces transferred there. The second of (3) shows that the motion about G is in a sense independent of its own instantaneous motion (but the unexplicit force terms may conceal an actual 'coupling'). In particular applications the more flexible (4) has decided advantages, as we shall find.

When a continuum is *rigid*, the 6 component equations of motion for the whole body correspond exactly with its degrees of freedom, 3 translational and 3 rotational. This preliminary observation makes plausible a basic theorem proved rigorously in §3.2: the motion of a rigid body is uniquely determined by (3) when the overall resultants are given at all times together with initial data for position and velocity. In particular, a rigid body can remain at rest in a frame where the overall force and couple resultants vanish for an appropriate position and orientation in that frame. Conversely, these resultants necessarily vanish for a frame in which a continuum is in fact at rest. For a rigid body such a frame can always be found: it is specified by any triad of directions embedded in the material itself, relative to which the entire body is *ipso facto* motionless. In this limited sense the equations of motion of a single rigid body can always be brought to a statical form.† On the other hand, we must beware of concluding that Euler's postulate is implied by the equations of pure statics, considered to be established by observations of forces on bodies at rest in the earth's frame alone. For, while this frame is admittedly typical, the total situation is not, since the *actual* forces (spring, contact, gravitational) are here virtually in balance (i.e., aside from the small centrifugal body-force, which anyway vanishes at the poles); whereas Euler's equations are an assertion about general situations in which the *actual* forces can be arbitrarily out of balance. This is clearly a real physical distinction, not removable by mere 'book-keeping' with fictitious body-forces.

When the continuum is *non-rigid*, its degrees of freedom are infinite. Even when applied in detail to arbitrary parts, the equations of balance still do not fix the motion relative to G. On the other hand, they do have definite implications for the internal tractions (§2.11). But for the present we confine attention to entire bodies and their overall properties. Two **conservation principles** are immediate:

(i) In a frame where the associated force resultant vanishes identically in time the linear momentum does not vary and the mass-centre moves uniformly or remains at rest.

† This at least supplies a non-trivial interpretation of one version of 'd'Alembert's principle', by which all mass-accelerations are formally transferred to the other side of the equations of motion and their negatives called 'inertia forces'.

(ii) In a frame where the associated couple resultant at G, or at any fixed point P, vanishes identically in time the corresponding moment of momentum does not vary.

More especially, these principles hold also in respect of any single component of $\mathbf{v}_G$ or $\mathbf{h}$ on a direction through G or P fixed in the considered frame, provided the same component of $\mathbf{R}$ or $\mathbf{M}$ vanishes perpetually. It need hardly be remarked that the momenta relative to any other frame will not usually be conserved, since the apparent velocities are different. The general connexions are obtained later but we can notice here that, in all frames with identical spins, the motion of the body relative to G appears the same and hence so does $\mathbf{h}(G)$; correspondingly the translational body-forces, though different, are always uniform and do not contribute to $\mathbf{M}(G)$.

It is of course always possible to find at least one frame relative to which the overall momenta both have assigned constant values under any actual forces. But this possibility only becomes of interest when the frame in question has a sufficiently simple and determinate motion against the sidereal background, so that the corresponding body-forces are themselves simple functions of space and time. For example, in a situation where (15) in §2.9 applies, $\mathbf{h}(G)$ is conserved when the Coriolis couple is exactly annulled by the couple resultant of the surface tractions; in the earth's frame, say, the needed resultant is very small. Again, in any sidereal frame, $\mathbf{h}(G)$ is *effectively* constant for a body of ordinary dimensions under no traction couple, in so far as the gravitational field of the solar system can be treated as locally uniform. The moment of momentum is *precisely* constant in a centrobaric body, for which the resultant gravitational action of any external mass-element (and hence all other matter) is a pure force directed through G irrespective of orientation (§1.16).

With typical celestial bodies, on the other hand, the individual moments of momentum are not conserved in a sidereal frame. The mutual gravitational actions between any pair are not single forces in the line of their mass-centres (see §§1.16 and 3.10). But, since the actions are pairwise in balance, the whole moment of momentum of the solar system about its own mass-centre C is conserved (or about any point fixed in a Newtonian frame, where C moves uniformly). In detail let $\mathbf{r}_P, \mathbf{r}_Q, \ldots$ be the positions of the individual mass-centres $P, Q, \ldots$ with respect to C. Let $(\mathbf{R}_{PQ}, \mathbf{M}_{PQ})$ be the gravitational force and moment resultants at P due to body Q, and $(\mathbf{R}_{QP}, \mathbf{M}_{QP})$ the balancing resultants at Q due to body P; then

$$\mathbf{R}_{PQ} + \mathbf{R}_{QP} = \mathbf{0}, \quad \mathbf{M}_{PQ} + \mathbf{M}_{QP} = (\mathbf{r}_Q - \mathbf{r}_P) \wedge \mathbf{R}_{PQ}. \tag{5}$$

In a Newtonian frame of reference

$$(m\mathbf{a})_P = \sum_{Q \neq P} \mathbf{R}_{PQ}, \quad \frac{d}{dt}\mathbf{h}(P) = \sum_{Q \neq P} \mathbf{M}_{PQ}, \tag{6}$$

by applying (3) to a typical body, where $\mathbf{a}_P$ is the acceleration of P. The 'orbital' moment of momentum of the system relative to C changes at the rate

$$\sum_P (\mathbf{r} \wedge m\mathbf{a})_P = \sum_{\text{pairs}} (\mathbf{r}_P - \mathbf{r}_Q) \wedge \mathbf{R}_{PQ}$$

from the first of (5) and (6). The whole 'rotational' moment of momentum, by definition the sum of the individual momenta about the mass-centres, changes at the rate

$$\sum_P \frac{d}{dt}\mathbf{h}(P) = \sum_{\text{pairs}} (\mathbf{r}_Q - \mathbf{r}_P) \wedge \mathbf{R}_{PQ}$$

from the second of (5) and (6). Consequently, the orbital and rotational momenta may vary separately when each $\mathbf{R}_{PQ}$ is not at all times parallel to PQ, but their sum remains constant (cf. §1.13). And this sum is of course the conserved moment of momentum of the system about C, by (6) in §2.9 applied to all the bodies separately.

2.11 Stress Analysis in a Continuum

We explore here the implications of the Eulerian postulates in regard to the field of traction inside a continuum. These fundamental results in stress analysis are mainly due to Cauchy (1822), though special cases were anticipated by Euler. While scarcely needed in what follows, they are included for the perspective they bring to rigid-body dynamics itself.

It is remarkable how much flows from the linear equations of motion alone:

$$\int(\dot{\mathbf{v}} - \boldsymbol{\gamma})dm = \int \boldsymbol{\tau} dS \tag{1}$$

for any part of a body (written in this form, both sides are invariant under change of frame). First we notice an immediate corollary:

the mutual tractions of the regions on either side of an imagined interface are locally equal and opposite, the acceleration being finite though not necessarily continuous.† This is seen by applying (1) to a vanishingly thin disc of material based on an arbitrary *fixed* element of interface; in the limit the mass integral vanishes but the surface integral approaches the product of the fixed basal area and the supposed difference in mutual tractions, which difference must

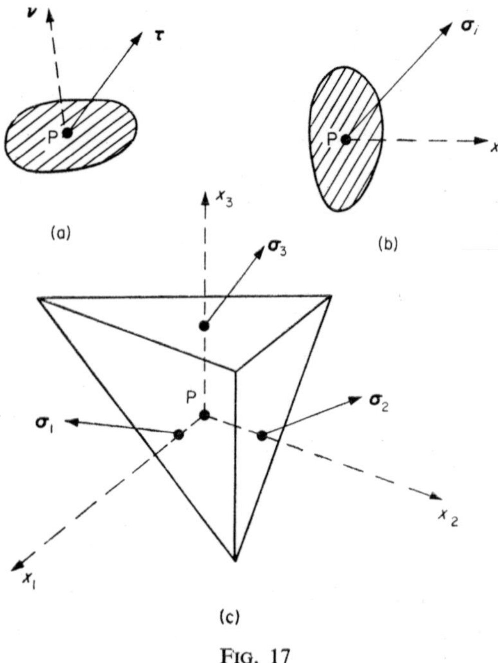

FIG. 17

therefore itself be zero. Conversely, from Newton's second law for an infinitesimal volume element (mass-point), together with the third law postulated in respect of internal tractions, it can be deduced that (1) holds for any finite region. For, in a Riemann sum over an arbitrary partition (in limiting replacement of the volume integral) the equal and opposite traction contributions cancel over the internal

† It is possible, at least as an idealization, to admit isolated surfaces (shocks) where the velocity of a mass element changes abruptly as it crosses the surface. There is then a corresponding jump in interfacial traction.

boundaries of the elementary cells, leaving only the integral over the external surface of the whole region.

Let the unit normal at a generic point P of an interface be denoted by $\boldsymbol{\nu}$, its sense being chosen arbitrarily. Let $\boldsymbol{\tau}$ denote the local traction exerted *by* the material at the side to which $\boldsymbol{\nu}$ points (Fig. 17a), while $-\boldsymbol{\tau}$ is the traction acting *on* that material. We show that $\boldsymbol{\tau}$ depends *linearly* on $\boldsymbol{\nu}$ for varying orientations of the interface at P. Choose orthogonal directions, typically x_i ($i = 1, 2, 3$), and let $\boldsymbol{\sigma}_i$ be the traction at P on an interface locally perpendicular to x_i (Fig. 17b). Consider a vanishingly small tetrahedron of material containing P and such that $\boldsymbol{\nu}$ is the outward normal to one face while the other faces are parallel to the coordinate planes. With the configuration of Fig. 17c the respective facial areas are in the ratios $1 : -\nu_i$. Equation (1) applies to a tetrahedron with volume ε^3, where ε is a linear dimension, then gives†

$$(\boldsymbol{\tau} - \nu_i \boldsymbol{\sigma}_i)\varepsilon^2 = O(\varepsilon^3);$$

the same expression is obtained when $\boldsymbol{\tau}$ is directed into any other octant, having regard to the sign convention for traction. Since ε can be made arbitrarily small it necessarily follows that, at P,

$$\boldsymbol{\tau} = \nu_i \boldsymbol{\sigma}_i \text{ or } \tau_j = \sigma_{ij}\nu_i \quad (j = 1, 2, 3) \tag{2}$$

where σ_{ij} denotes the jth component of $\boldsymbol{\sigma}_i$. This establishes the stated linearity. The nine σ_{ij} ($i, j = 1, 2, 3$) collectively specify the state of stress at P, and are called *stress components*. Their connexion with the set of components for any other triad is obtained later. Meanwhile we observe that the state of stress at a point is fully determined by just one such set, in that the traction on any surface element is calculable from (2).

If the surface integral over the tetrahedron were evaluated to order ε^3, the coefficient would plainly have the interpretation of resultant force per unit volume due to interfacial tractions. But (1) implies its independence of the actual shape of the infinitesimal element, by asserting its equality with $\rho(\dot{\mathbf{v}} - \boldsymbol{\gamma})$ where ρ is the mass density at P. For simplicity we consider instead an infinitesimal

† Summation convention. A repeated literal subscript means that the term is summed over all possible numerical values of that subscript; for example,

$$a_i b_i \equiv a_1 b_1 + a_2 b_2 + a_3 b_3, \qquad c_{ij} x_j \equiv c_{i1} x_1 + c_{i2} x_2 + c_{i3} x_3.$$

cube, of volume ε^3 say, with edges parallel to the coordinate directions. The net contribution to the jth component of the traction integral from the faces perpendicular to x_1 is $(\partial\sigma_{1j}/\partial x_1)\varepsilon \times \varepsilon^2$ to third order, or $\partial\sigma_{1j}/\partial x_1$ per unit volume in the limit. With similar contributions from the other faces we have, in all,

$$\rho(\dot{v}_j - \gamma_j) = \frac{\partial\sigma_{ij}}{\partial x_i} \quad (j = 1, 2, 3). \tag{3}$$

This is Cauchy's form of the linear equations of motion. Conversely, suppose that the stress components are given as functions of position, piecewise continuous subject to the continuity of traction (2), and such that the stress gradients exist and satisfy (3). We prove that (1) is necessarily satisfied with respect to *any* regular region.

By hypothesis this consists of a finite number of sub-regions in each of which the components of stress are continuous. By applying the divergence theorem to each sub-region and cancelling the equal and opposite contributions from their mutual boundaries, one has for the whole volume

$$\int \frac{\partial\sigma_{ij}}{\partial x_i}\frac{dm}{\rho} = \int \sigma_{ij}\nu_i dS$$

for each value of the free subscript j (dm/ρ being the element of volume). But by (2) and (3) these expressions are equal to the respective sides of (1). (The proof becomes intuitive on reflecting that any sub-region can be approximated by indefinite partition into cubes and tetrahedra.) This also establishes the equivalence of two ways of formulating the balance of force and mass-acceleration: one global (via overall integrals over arbitrary regions) and the other local (via differential relations and boundary conditions at internal or external surfaces).

It remains to discover what is implied by the rotational equation of balance:

$$\int \mathbf{r} \wedge (\dot{\mathbf{v}} - \boldsymbol{\gamma})dm = \int \mathbf{r} \wedge \boldsymbol{\tau} dS \tag{4}$$

where, to be definite, moments are taken about the origin of coordinates. From (2) the components of the surface integral are expressible as

$$\int (x_j\tau_k - x_k\tau_j)dS = \int (x_j\sigma_{ik} - x_k\sigma_{ij})\nu_i dS$$

where $(j, k) = (1, 2)$ or $(2,3)$ or $(3,1)$, the axes being right-handed. By the divergence theorem and continuity of traction this becomes

$$\int \frac{\partial}{\partial x_i}(x_j\sigma_{ik} - x_k\sigma_{ij})\frac{dm}{\rho}$$

and finally

$$\int\left[\left(x_j\frac{\partial\sigma_{ik}}{\partial x_i} - x_k\frac{\partial\sigma_{ij}}{\partial x_i}\right) + (\sigma_{jk} - \sigma_{kj})\right]\frac{dm}{\rho}$$

with the help of $\partial x_j/\partial x_i = \delta_{ij}$ (Kronecker delta), $\delta_{ij}\sigma_{ik} = \sigma_{jk}$, etc. The first bracket in the integrand is recognizably the contribution to the moment from the resultant of the tractions over a typical element; consequently, the second bracket must be the contribution from the moment of these tractions about the centroid of the element. Since the region is arbitrary the ith component of centroidal moment per unit volume has the unique value

$$\sigma_{jk} - \sigma_{kj} \quad (i \neq j \neq k, \text{ cyclic order}). \tag{5}$$

But, by (3), the left side of (4) is balanced by the former contributions alone, and (5) must vanish at every point:†

$$\sigma_{jk} = \sigma_{kj} \quad (\text{all } j, k). \tag{6}$$

Conversely either (1) with (6), or (2) and (3) with (6), imply (5) for arbitrary regions.

The symmetry of the 3×3 matrix array of stress components (proved by Cauchy) may be regarded as the essential content of Euler's postulates for a continuum, over and above what can be brought within the framework of Newton's laws. By contrast, in a discrete system of idealized mass-points, the Newtonian laws in themselves imply the rotational equation of balance. For, if $m\dot{\mathbf{v}} = \mathbf{F}$ for the typical mass-point $P(\mathbf{r})$ under a resultant force $\mathbf{F}$,

$$\Sigma(\mathbf{r} \wedge m\dot{\mathbf{v}}) = \Sigma(\mathbf{r} \wedge \mathbf{F})$$

for any sub-system, simply by multiplying each individual equation vectorially by the corresponding $\mathbf{r}$ and summing. On the right-hand

† The possibility of *body-moments* has been disregarded in (4). They occur, for instance, in a polarized medium under an electric or magnetic field. In such cases the difference (5) is clearly equal to the ith component of reversed body-moment per unit volume.

side there is a pairwise cancellation of contributions from certain mutual interactions which are equal and opposite and in line; in particular the gravitational forces are of this type when the body dimensions are vanishingly small (cf. §2.10). The moment of mass-accelerations is therefore automatically equal to the moment of the remaining forces only, and no separate postulate is needed. (The remaining actual forces are purely external when all mutual forces cancel; this does not always happen, for instance with a system of charged particles moving in their own electromagnetic fields.)

A corollary of the symmetry property is worth notice. The tractions $\boldsymbol{\tau}(\boldsymbol{\nu})$ and $\boldsymbol{\tau}^*(\boldsymbol{\nu}^*)$ over any pair of plane elements at a given point have equal components $\boldsymbol{\tau}\boldsymbol{\nu}^*$ and $\boldsymbol{\tau}^*\boldsymbol{\nu}$ on the other's normal. For

$$\tau_j \nu_j^* = (\sigma_{ij}\nu_i)\nu_j^* = (\sigma_{ji}\nu_j^*)\nu_i = \tau_i^*\nu_i.$$

In particular, (6) itself is a statement that the component of $\boldsymbol{\sigma}_j$ on the x_k direction is equal to the component of $\boldsymbol{\sigma}_k$ on the x_j direction. Incidentally, the right side of (3), which is the jth component of force per unit volume due to internal traction, can be written as $\partial\sigma_{ji}/\partial x_i$ since $\sigma_{ij} = \sigma_{ji}$ and so as div $\boldsymbol{\sigma}_j$.

Finally, the transformation law for stress is derived from (2). Take any other orthogonal triad specified by unit vectors $\mathbf{l}_p(p = 1, 2, 3)$, so that l_{pi} is the cosine of the angle between the pth new axis and the ith old one. Let $\boldsymbol{\tau}_p$ be the traction over the plane element with normal $\mathbf{l}_p$. Then the pq component of stress with respect to the new axes is

$$\sigma_{pq}^* = \mathbf{l}_q\boldsymbol{\tau}_p,$$

by simple resolution. But the components of the vectors in this scalar product are l_{qj} and $l_{pi}\sigma_{ij}$ on the old axes ($j = 1, 2, 3$). Consequently

$$\sigma_{pq}^* = l_{pi}l_{qj}\sigma_{ij}, \tag{7}$$

with summation as usual over both repeated subscripts. This is precisely the transformation property characterizing any *cartesian tensor* of second rank. Such a tensor is the totality of sets of 3 × 3 components which are associable with all orthogonal axes by some definite rule (depending on the physical or geometrical context) and which transform like the set of components $a_i b_j$ where **a** and **b** are any vectors:

$$a_p^* = l_{pi}a_i \equiv \mathbf{l}_p\mathbf{a},$$
$$b_q^* = l_{qj}b_j \equiv \mathbf{l}_q\mathbf{b},$$
$$(a_p^* b_q^*) = l_{pi}l_{qj}(a_i b_j).$$

We shall later encounter another example of a second rank tensor (§3.4).

This concludes the present brief analysis of stress properties common to any material. In order to determine the actual field of stress in a specified material the particular stress/strain properties must naturally be adjoined to the equations of balance (2), (3) and (6).†

2.12 Principles of Work and Energy

The work-rate or power of a force, actual or fictitious, acting on a mass-point P is by definition its scalar product with the current velocity of P.†† Since velocity depends on the frame of reference so does the work-rate of an actual, invariant, force. For example, the contact force at a frictionless and non-deforming constraint generally does work, but not in the particular frame relative to which the constraint is at rest (§2.3). Again, no force does work when the velocity vanishes, which is trivially always the case in a frame with P as origin. However, in all frames of reference, it is true that the work-rate of the force resultant on P is equal to the rate of change of its apparent kinetic energy:

$$\frac{dT}{dt} \equiv \frac{d}{dt}\left(\tfrac{1}{2}m\mathbf{v}^2\right) = m\mathbf{v}\dot{\mathbf{v}} = \mathbf{F}\mathbf{v}. \tag{1}$$

There may be contributions by some of the fictitious body-forces associated with the adopted frame but never from the Coriolis force, which is always perpendicular to the apparent velocity.

† For instance, in a perfect fluid σ_{ij} is by definition always of form $-p\delta_{ij}$ where p is a positive scalar, known as the pressure, usually related in a prescribed way to the density.

For a synoptic account of modern developments in various branches of solid or fluid mechanics see W. Prager: *Introduction to Mechanics of Continua* (Ginn, 1961).

†† The c.g.s. unit of work is an *erg* and corresponds to a force of 1 dyne acting through 1 cm. The m.k.s. unit is a *joule*, or newton-metre, equal to 10^7 ergs. In both systems the unit of work-rate is a *watt*, equal to 1 joule/sec or 10^7 dyne cm/sec.

Several types of force can be expressed as the (negative) gradient of a scalar potential. Principal among these are the gravitational actions of other matter (§1.14); the centrifugal body-force which admits the axisymmetric potential

$$-\tfrac{1}{2}m(\mathbf{\Omega} \wedge \mathbf{r})^2,$$

and the translational body-force which admits the plane potential

$$m\boldsymbol{\alpha}_O\mathbf{r}.$$

The Euler body-force is not a gradient since its curl is $-2m\dot{\mathbf{\Omega}}$ and only vanishes with the force itself (§2.8). These particular potentials may depend explicitly on time as well as on position $\mathbf{r}$; respectively through changes in the configuration of the attracting matter, through variations in the spin $\mathbf{\Omega}$ of the adopted frame against a sidereal background, and through variations in the acceleıation $\boldsymbol{\alpha}_O$ of its origin. Generally, let $V(\mathbf{r}, t)$ denote the collective potential of these and any other gradient forces present in a considered situation, and write

$$\mathbf{F} = \boldsymbol{\mathcal{F}} - \operatorname{grad} V$$

where $\boldsymbol{\mathcal{F}}$ is the remaining resultant. The total rate of change of V, following the mass-point, is

$$\frac{dV}{dt} = \frac{\partial V}{\partial t} + \mathbf{v} \operatorname{grad} V$$

and so

$$\frac{d}{dt}(T + V) = \boldsymbol{\mathcal{F}}\mathbf{v} + \frac{\partial V}{\partial t}. \tag{2}$$

It may happen that a special frame can be found in which the potential, V^* say, depends only on the coordinates there so that $\partial V^*/\partial t = 0$. The corresponding gradient force is said to be **conservative**; the work done by it in passing from one position to another is equal to the decrease in V^*, independently of the intervening motion, and in particular vanishes over any closed circuit. If, also, the work-rate $\boldsymbol{\mathcal{F}}^*\mathbf{v}^*$ of the remaining force vanishes in this frame, an immediate first integral of the equations of motion is

$$T^* + V^* = \text{constant}, \tag{3}$$

in terms of the special coordinates. In these circumstances V^* can be said to be the 'potential energy' of the mass-point.

It is instructive to re-interpret, from this standpoint, some previous results for gravitating systems of free mass-points in a uniform external field. In a sidereal frame translating with the combined mass-centre the force potential for each mass is time-dependent, since the configuration varies. Consequently, in this frame, there are no integrals (3) for the masses separately; nor, in general, are there in any other frame, by the Bruns–Painlevé theorem (§1.10). But, as we know, there is such an integral for the system as a whole since the overall work-rate is the negative rate of change of a function of the configuration only [§1.10, equation (1)]. In the restricted problem of 3 bodies, however, Jacobi's integral is of type (3) for the infinitesimal mass P separately. V^*, which is $-mU$ in the notation of §1.9, consists of the gravitational and centrifugal potentials, neither of which depends explicitly on time in the frame attached to the supposedly rigidly-rotating pair of large masses; there is no Euler force, and the translational and external forces cancel (which is just another way of saying that the additional accelerations of P and the origin are presumed equal under the external field). The overall conservation integral in a sidereal frame is trivially satisfied here since, by hypothesis, the contribution from P is vanishingly small and from the others is constant. Similarly, for a mass-point under frictionless constraint fixed to the earth, V^* consists of the centrifugal and terrestrial potentials, written as $m\Phi$ in (5) of §2.3; the Euler force vanishes, the translational force once again cancels (very nearly) with the action of other matter, and the contact force does no work in the earth's frame.

In a continuum the kinetic energy changes at the rate

$$\frac{dT}{dt} = \int \mathbf{v}\dot{\mathbf{v}}dm = \int \boldsymbol{\gamma}\mathbf{v}dm + \int v_j \frac{\partial \sigma_{ij}}{\partial x_i}\frac{dm}{\rho},$$

by the linear equation of balance, (3) in §2.11. On the other hand, the work-rate of the acting forces is

$$\int \boldsymbol{\gamma}\mathbf{v}dm + \int \boldsymbol{\tau}\mathbf{v}dS = \int \boldsymbol{\gamma}\mathbf{v}dm + \int \frac{\partial}{\partial x_i}(\sigma_{ij}v_j)\frac{dm}{\rho},$$

using first (2) in §2.11 and then the divergence theorem. The preceding expressions differ by

$$\int\left[\frac{\partial}{\partial x_i}(\sigma_{ij}v_j) - v_j\frac{\partial\sigma_{ij}}{\partial x_i}\right]\frac{dm}{\rho} = \int\sigma_{ij}\frac{\partial v_j}{\partial x_i}\frac{dm}{\rho} = W\text{, say.} \tag{4}$$

Finally, then,

$$\frac{dT}{dt} + W = \int \boldsymbol{\gamma}\mathbf{v}dm + \int \boldsymbol{\tau}\mathbf{v}dS \tag{5}$$

is the work principle for a continuum.

To interpret the quantity denoted by W, imagine the region partitioned into infinitesimal elements. The work-rate of the tractions on the external surface is also the net work-rate of the tractions on the entire boundaries of the individual elements (since there is a cancellation at the mutual interfaces). But the contribution from an element is, in turn, equal to the work-rate of the associated tractions in its centroidal motion, amounting to $v_j\partial\sigma_{ij}/\partial x_i$ per unit volume, together with the work-rate of these tractions in the motion relative to the centroid; which last, by (4), can be none other than $\sigma_{ij}\partial v_j/\partial x_i$ per unit volume. Now, if the material were rigid (even momentarily), the matrix $\partial v_j/\partial x_i$ would be antisymmetric at every point: $\partial v_j/\partial x_i = -\partial v_i/\partial x_j$ for any pair i, j.† But by the rotational equation of balance the stress σ_{ij} is symmetric, (6) in §2.11. Consequently the product $\sigma_{ij}\partial v_j/\partial x_i$ summed over all i, j would vanish identically, since it consists of pairs of terms in which one factor is the same while the other is merely reversed in sign. The conclusion is that the integrand in W is zero where the material is locally rigid, and, conversely, the material cannot be rigid where the integrand is non-zero. W is thereby identified as the net work-rate in internal distortion. And, for a rigid body, the principle (5) reduces to the statement that the kinetic energy increases at the rate at which the overall forces do work. Another proof is given later in §3.2 without reference to the internal field of stress.

† Rigidity means that there is an unvarying distance between any pair of points embedded in the material. Otherwise expressed, the instantaneous rate of change of length of any line element $\delta\mathbf{r} \equiv \delta x_i$ vanishes. Equivalently, the relative velocity of its ends must be perpendicular to its length, or $(\delta x_j \partial v_i/\partial x_j)\delta x_i = 0$. At any considered station this quadratic must vanish for all possible directions $\delta\mathbf{r}$, which is so if and only if its matrix is antisymmetric.

The kinetic energy of a continuum may be decomposed into a translational (or orbital) part, associated with the motion of the mass-centre, and a rotational (or spin) part, associated with the motion relative to the mass-centre:

$$T = T_{\text{trans}} + T_{\text{rot}} = \tfrac{1}{2}m\mathbf{v}_G^2 + \tfrac{1}{2}\int(\mathbf{v} - \mathbf{v}_G)^2 dm. \tag{6}$$

This was first noticed by König (1751) for a system of mass-points. The proof is immediate: in the mass integral of the identity

$$\tfrac{1}{2}(\mathbf{v}^2 - \mathbf{v}_G^2) \equiv \tfrac{1}{2}(\mathbf{v} - \mathbf{v}_G)^2 + \mathbf{v}_G(\mathbf{v} - \mathbf{v}_G)$$

the final term vanishes through the now familiar property (2) in §2.9. The rate of change of the translational energy is

$$\frac{dT_{\text{trans}}}{dt} = m\mathbf{v}_G\dot{\mathbf{v}}_G = \mathbf{R}\mathbf{v}_G \tag{7}$$

in analogy to (1), where $\mathbf{R}$ is the overall force resultant defined previously. Whence, by subtraction from (5),

$$\frac{dT_{\text{rot}}}{dt} + W = \int\boldsymbol{\gamma}(\mathbf{v} - \mathbf{v}_G)dm + \int\boldsymbol{\tau}(\mathbf{v} - \mathbf{v}_G)dS. \tag{8}$$

On the right appears the work-rate in the motion relative to the mass-centre. In fact, the equation can be read as the work principle formulated in a frame translating with the mass-centre (the extra translational body-forces do no work overall).

In ordinary circumstances where (15) in §2.9 is valid, $\boldsymbol{\gamma}$ can be replaced in the work-rate by gravity $\mathbf{g}$ and treated as constant (the Coriolis component makes no contribution). Then (8) reduces to

$$\frac{dT_{\text{rot}}}{dt} + W = \int\boldsymbol{\tau}(\mathbf{v} - \mathbf{v}_G)dS \tag{9}$$

while (7) can be re-arranged as

$$\frac{d}{dt}(T_{\text{trans}} - m\mathbf{g}\mathbf{r}_G) = \left(\int\boldsymbol{\tau}dS\right)\mathbf{v}_G \tag{10}$$

in analogy to (2) with $\partial\mathbf{g}/\partial t = \mathbf{0}$.

For an effectively isolated system of extended bodies $\boldsymbol{\gamma}$ is equal to the resultant field $\mathbf{f}$ of the system, when the adopted frame is sidereal and translates with the combined mass-centre. Hence, by (5) in §1.14 applied to the incremental displacements in an actual motion, the overall work-rate of the body-forces is $-dV/dt$ where V is the gravitational potential energy of the system. Consequently, if there are no surface tractions,

$$\frac{d}{dt}(T + V) = -W,$$

where T and W now denote the total kinetic energy and distortion work-rate, and either gravitational or inertial mass is used consistently throughout. For a system of actual celestial bodies W can be neglected over reasonable periods of time and then $T + V$ is constant (cf. §1.13). However, over long periods, a decrease in rotational energy through the presence of bodily or oceanic tides becomes noticeable.

EXERCISES

1. By direct calculation verify the invariant property of additional acceleration (§2.1) in the following circumstances. When $t \leqslant 0$ a particle describes a circle of radius a in a frame β at uniform angular rate Ω; when $t \geqslant 0$ it moves along a fixed tangent with speed $a\Omega$. Describe qualitatively the apparent path for all t in a frame α rotating uniformly at rate Ω, in the same sense relative to β, about an axis perpendicular to the plane of the circle and through its centre, and obtain the parametric equations

$$x = a(\cos\theta + \theta\sin\theta), \qquad y = a(-\sin\theta + \theta\cos\theta),$$

when $\theta = \Omega t \geqslant 0$. Show, in particular, that the apparent path touches the x axis and has locally infinite curvature, while at $t = 0$ the velocity is continuous but the acceleration jumps by the amount $a\Omega^2$ in the x direction.

2. An artificial satellite S is in free circular orbit about the centre E of the earth. Perturbations by other matter are neglected and the earth is regarded as centro-symmetric. Axes (ξ, η) are taken in the orbital plane with origin at S and ξ outwards along ES. Obtain expressions for the components of the earth's field near S, and show (cf. §2.1) that the equations of motion of a nearby mass-point P in the same plane are

$$\ddot{\xi} - 2\omega\dot{\eta} - 3\omega^2\xi = a_\xi, \qquad \ddot{\eta} + 2\omega\dot{\xi} = a_\eta,$$

to first order, where $2\pi/\omega$ is the sidereal period of the orbit and (a_ξ, a_η) denotes the additional acceleration of P through causes other than the earth's field.

3. Derive the path of a missile in free motion, to the approximation involved in (4) of §2.2:

$$x - ut = \Omega vt^2 \sin\lambda,$$
$$y - vt = \Omega(\tfrac{1}{3}gt\cos\lambda - u\sin\lambda - w\cos\lambda)t^2,$$
$$z - wt = (\Omega v\cos\lambda - \tfrac{1}{2}g)t^2,$$

where (u, v, w) is the initial velocity and the (x, y, z) axes are based at the firing point in latitude λ with x south, y east, and z vertical (as in Fig. 11). By considering the sign of $vx - uy$, or otherwise, show that at the point of fall (regarded as in the plane $z = 0$) the deviation from the plane of projection is to the right or left according as $3(u^2 + v^2) \tan \lambda + uw$ is positive or negative. Obtain formulae, correct to first order in Ω, for the effects of the earth's rotation on the time of flight and the range; investigate the magnitudes of these effects in typical cases.

4. A particle is smoothly constrained to move on a plane which is inclined at α to the horizontal and rotated about the vertical with constant spin ω. Disregarding the earth's rotation, obtain the equations of motion in the frame of reference turning with the plane:

$$\ddot{x} - 2\omega\dot{y}\cos\alpha - \omega^2 x = 0,$$
$$\ddot{y} + 2\omega\dot{x}\cos\alpha - \omega^2 y \cos^2\alpha = 0,$$

where the (x, y) axes are fixed in the plane, with origin at the possible position of relative rest and y directed up the line of greatest slope. [It may be found convenient to take auxiliary axes as in (4) of §2.1 and then resolve the contributions (i) and (iii) in the x, y directions.]

Show how the general solution can be constructed from particular integrals of type

$$x = a \exp(\lambda\omega t), \qquad y = b \exp(\lambda\omega t),$$

where

$$\lambda^4 + (3\cos^2\alpha - 1)\lambda^2 + \cos^2\alpha = 0.$$

When $\cos\alpha < \frac{1}{3}$ deduce that the path is always asymptotically straight as $\omega t \to \infty$.

5. Consider motion under axisymmetric constraint (§2.5) from the frame of reference turning round the axis of symmetry with the meridian plane through the particle. Detail the appropriate centrifugal, Coriolis and Euler body-forces. Explain why only the first does work in this frame and show that it admits the potential $h^2/2r^3$ per unit mass. Hence formulate the work principle in the moving frame (cf. §2.12) and re-derive the first integrals (3) and (5) in §2.5.

6. For a spherical pendulum, and with the notation of §2.5, show that the inclination θ of the string to the downward vertical satisfies

$$\frac{1}{2}\left[\frac{d}{dt}(\cos\theta)\right]^2 = \frac{g}{l}(\cos\alpha - \cos\theta)(\cos\theta - \cos\beta)\left(\cos\theta + \frac{1 + \cos\alpha\cos\beta}{\cos\alpha + \cos\beta}\right).$$

Obtain a formula, in terms of an integral in the variable $\cos\theta$, for the azimuthal angle between successive points of tangency at the levels $\theta = \alpha, \beta$.

7. A particle moves freely under gravity inside a fixed circular cone of angle 2α with its axis vertical. Deduce from the conservation integrals in §2.5 that, if the surface were developed on a plane, the resulting trace of the original orbit could be described under central acceleration $g\cos\alpha$ and constant moment of velocity $h/\sin\alpha$.

8. A particle is smoothly constrained to move on a rigid plane curve $z = f(x)$ (representing a wire or fine tube for example). This is rotated with constant spin ω about the z axis which is vertical and fixed to the earth (whose own spin is to be disregarded).

(i) Adopting the standpoint of an observer in the frame rotating with the constraint, show that the centrifugal body-force admits the potential $-\frac{1}{2}\omega^2x^2$ per unit mass and obtain the conservation integral

$$\tfrac{1}{2}u^2 + (gz - \tfrac{1}{2}\omega^2x^2) = \text{constant},$$

where u is the velocity relative to the constraint [cf. (3) in §2.12]. Reconcile this integral with the work principle in the terrestrial frame [cf. (2) in §2.12, with $\partial V/\partial t = 0$], showing that the circumferential component of the contact force does work in this frame at rate $d(\omega^2x^2)/dt$ per unit mass.

(ii) From the stability analysis in §2.5 with Ψ there replaced by $gz - \frac{1}{2}\omega^2x^2$ prove that, if x_0 is a position of relative rest ($u = 0$), it is stable when

$$f''(x_0) > f'(x_0)/x_0.$$

State the frequency of small oscillations about such a position.

(iii) When the constraint is a circle of radius a, establish that the (lowest) position of equilibrium is stable if and only if $a\omega^2 \leqslant g$. For this purpose expand the total potential in powers of x and consider the sign of its increment.

(iv) Discuss the stability of all positions of relative rest on constraints defined respectively by $a^2z = \frac{1}{3}x^3 + \frac{1}{2}ax^2$ and $2a^2z = \frac{1}{3}x^3 + a^2x$.

(v) Assess the stability of the two distinct positions of relative rest on the constraint $z/a = 1 - \cos(x/a)$, $|x/a| \leqslant \pi$, when $a\omega^2 < g$, using only graphical arguments. Examine separately the critical case $a\omega^2 = g$.

9. A particle is smoothly constrained to move on a plane rigid curve which is rotated in a horizontal plane at constant rate ω about a fixed vertical axis.

(i) Take the standpoint of an observer in the frame rotating with the constraint and follow the procedure indicated in the previous question. Prove that

$$u^2 - \omega^2r^2 = \text{constant},$$

where r is the radius from the axis and u is the velocity relative to the constraint.

(ii) Obtain the contact force per unit mass in the form

$$\frac{u}{r}\frac{d}{dr}(pu) + 2\omega u$$

where pu is the moment of relative velocity about the axis. (Recall the standard formula u^2dp/rdr for the normal component of acceleration, in the adopted frame.) Reduce the above expression for a motion initiated from rest in the terrestrial frame in a position $p = r$.

(iii) Show that positions of relative rest are possible only where the radius has a stationary value, or at the origin if the curve happens to intersect the axis. By considering the behaviour of r^2 near such positions, preferably geometrically, show that there is instability when the local radius of curvature exceeds r. (In an analytical treatment note that r itself is unsuitable as independent variable and use the arc-length.)

10. Three frames of reference α, α' and β have a common axis and origin, while frames α and α' have uniform spins Ω and Ω' relative to β. The centrifugal and Coriolis body-forces for α *vis-à-vis* β are denoted by $\mathbf{P}$ and $\mathbf{Q}$ respectively (§2.8). Show that the total fictitious body-force for α' *vis-à-vis* α is expressible as

$$n(\mathbf{Q} - n\mathbf{P}) \qquad \text{where} \quad \Omega' = \Omega(1 + n).$$

Find also the fictitious body-force for α' *vis-à-vis* β and check that it is otherwise obtainable simply by adding $\mathbf{P} + \mathbf{Q}$ to the above expression.

CHAPTER III

GENERAL MOTION OF A RIGID BODY

3.1 Kinematics

In rigid-body mechanics we study the implications of Euler's equations of balance (§2.10) under the supposition that no part of the continuum may deform. The body can then be treated exclusively as a whole, without reference to the internal state of stress. In this respect the idealization is admissible as an approximation for any actual body that deforms only infinitesimally in a considered motion, say with fractional changes in dimensions of order not more than 10^{-4} to 10^{-3}. Such strains are, for example, typical for metals and other structural materials loaded within their elastic ranges. Additionally, the applied forces must not vary too rapidly in relation to the actual speed of propagation of stress waves;† otherwise, despite the smallness of the distortion, the distribution of velocity can be far different from that in a rigid body (where the wave speed is infinite). The idealized distribution will now be examined, in readiness for the eventual computation of moments of momentum and rotational energy.

A minimal characterization of rigidity in a continuum is that mutual distances between all points are invariant. Dependent properties follow by elementary geometry, for example that embedded lines or planes remain lines or planes and preserve their mutual inclinations.†† More especially, the entire configuration in relation to a considered particle P is determined by the orientations of a chosen pair of embedded radii at P, with unit vectors $\mathbf{l}_1$ and $\mathbf{l}_2$ say. Of their 6 components only 3 are independently variable, because of the connexions $\mathbf{l}_1^2 = 1 = \mathbf{l}_2^2$, $\mathbf{l}_1\mathbf{l}_2 =$ constant (in practice usually

† For an introductory account see H. Kolsky: *Stress Waves in Solids* (Clarendon Press 1953).

†† This property is found also in uniform expansion or contraction (similarity transformation) and is therefore insufficient by itself for rigidity.

zero, the directions being orthogonal). Correspondingly, their rates of change are subject to

$$\mathbf{l}_1\dot{\mathbf{l}}_1 = 0 = \mathbf{l}_2\dot{\mathbf{l}}_2, \quad \mathbf{l}_1\dot{\mathbf{l}}_2 + \dot{\mathbf{l}}_1\mathbf{l}_2 = 0.$$

These rates should be thought of as the velocities of particles at unit distance from P in positions $\mathbf{l}_1$ and $\mathbf{l}_2$.

We prove first a theorem of Euler, to the effect that the instantaneous motion of a rigid body relative to any of its particles is always a pure spin. Let $\mathbf{l}$ be a unit vector along any other embedded radius at P. From the invariance of the scalar products $\mathbf{l}_1\mathbf{l}$ and $\mathbf{l}_2\mathbf{l}$,

$$\mathbf{l}_1\dot{\mathbf{l}} = -\mathbf{l}\dot{\mathbf{l}}_1 \quad \text{and} \quad \mathbf{l}_2\dot{\mathbf{l}} = -\mathbf{l}\dot{\mathbf{l}}_2, \quad \text{with } \mathbf{l}\dot{\mathbf{l}} = 0.$$

These are to be read as a 3×3 system of linear equations uniquely determining the cartesian components of velocity $\dot{\mathbf{l}}$ provided the determinant or scalar triple product $\{\mathbf{l}_1\mathbf{l}_2\mathbf{l}\}$ does not vanish; that is, provided $\mathbf{l}$ is not coplanar with $\mathbf{l}_1$ and $\mathbf{l}_2$. In particular, when

$$\{\mathbf{l}_1\mathbf{l}_2\dot{\mathbf{l}}_1 \wedge \dot{\mathbf{l}}_2\} \equiv (\mathbf{l}_1\dot{\mathbf{l}}_2)^2 \equiv (\dot{\mathbf{l}}_1\mathbf{l}_2)^2 \neq 0,$$

the direction $\dot{\mathbf{l}}_1 \wedge \dot{\mathbf{l}}_2$ causes all the right sides to vanish identically and is therefore stationary in the adopted frame ($\dot{\mathbf{l}} = \mathbf{0}$). The motion about P can therefore only be a spin with this direction as axis. If, on the other hand, the above expressions are zero little remains to be proved. $\dot{\mathbf{l}}_1$ and $\dot{\mathbf{l}}_2$ are orthogonal to both $\mathbf{l}_1$ and $\mathbf{l}_2$ and hence parallel. Consequently, all particles in the plane through the reference directions have momentarily a purely normal motion relative to P, and this necessarily vanishes along a line through P (the velocity being a linear function of position since the plane remains plane).

The free vector whose magnitude is equal to the angular rate of rotation and which is directed parallel to the axis of rotation in a right-handed sense is called the body **spin.** Both its direction and magnitude generally vary in time. Of course, the spin is a property of the body as a whole, in no way tied to the choice of reference particle P. (Reflect that the axial direction is uniquely characterized as normal to the one family of embedded parallel planes whose orientation is not changing momentarily, and the spin magnitude as the rate of rotation either of lines fixed in such planes or of planes orthogonal to them.) Euler's result can now be expressed analytically: the distribution of relative velocity throughout a rigid body is

$$\left.\begin{aligned} \dot{\mathbf{r}} &= \boldsymbol{\omega} \wedge \mathbf{r} \\ \text{or equivalently} \qquad \dot{\mathbf{l}} &= \boldsymbol{\omega} \wedge \mathbf{l} \quad \text{with } \dot{r} = 0, \end{aligned}\right\} \tag{1}$$

where $\boldsymbol{\omega}(t)$ is the spin and $\mathbf{r} = r\mathbf{l}$ the relative position vector. The derivation of this simple but all-important formula is made clear in Fig. 18; the velocity of the particle at $\mathbf{r}$ in the rotation with respect

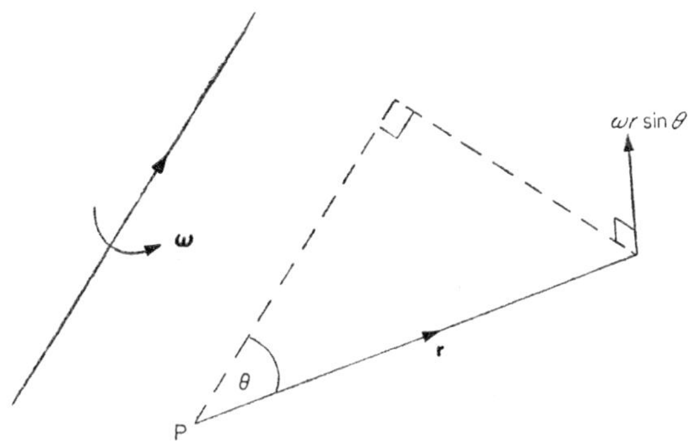

FIG. 18

to P is normal to the plane of $\boldsymbol{\omega}$ and $\mathbf{r}$, its magnitude is $\omega \times$ distance $r \sin\theta$ from the axis drawn through P parallel to $\boldsymbol{\omega}$, and its sense corresponds to the stated order of factors in the vector product.

An alternative proof of Euler's theorem is instructive. In the relative motion the velocities of particles on any given line through P are parallel and proportional to their distances from P, since a line neither lengthens nor distorts. In short, the distribution of relative velocity is homogeneous and linear, and in terms of cartesian axes in the adopted frame must be

$$\dot{x}_i = \omega_{ij} x_j \tag{2}$$

where ω_{ij} is some matrix function of time only. Moreover, for mutual distances to be preserved, we must have

$$\frac{d}{dt}(\tfrac{1}{2}x_i^2) = x_i \dot{x}_i = \omega_{ij} x_i x_j \equiv 0$$

identically. This requires the matrix to be antisymmetric:

$$\omega_{ij} = -\omega_{ji}$$

for all i, j. Since the determinant is therefore zero, being of odd order, a direction exists such that the right side of (2) vanishes; this direction is accordingly stationary and necessarily an axis of rotation, as before. To be specific: the matrix must be of form

$$\omega_{ij} \equiv \begin{pmatrix} 0 & -\omega_3 & \omega_2 \\ \omega_3 & 0 & -\omega_1 \\ -\omega_2 & \omega_1 & 0 \end{pmatrix} \tag{3}$$

where the independent elements $(\omega_1, \omega_2, \omega_3)$ define the axial direction. Correspondingly, relations (2) become $\dot{x}_1 = \omega_2 x_3 - \omega_3 x_2$, etc., and can be recognized as the components of (1) on the right-handed axes when $(\omega_1, \omega_2, \omega_3)$ are identified with the cartesian components of $\boldsymbol{\omega}$.

To see their kinematic significance imagine that the spin were to be decomposed vectorially into any number of arbitrarily oriented parts: $\boldsymbol{\omega} = \boldsymbol{\omega}' + \boldsymbol{\omega}'' + \ldots$. Then, by the distributive property of vector products, the motion (2) with respect to any considered particle P can plainly be regarded as the superposition of separate rotary motions $\boldsymbol{\omega}' \wedge \mathbf{r}$, $\boldsymbol{\omega}'' \wedge \mathbf{r}$, . . ., or in particular of separate spins $\omega_1, \omega_2, \omega_3$ about axes in the coordinate directions. Changing the viewpoint slightly and taking in illustration the simplest decomposition $\boldsymbol{\omega} = (\boldsymbol{\omega} - \boldsymbol{\Omega}) + \boldsymbol{\Omega}$, we can think of $\boldsymbol{\omega}' = \boldsymbol{\omega} - \boldsymbol{\Omega}$ as the spin of the body (or of any embedded triad) seen from an auxiliary frame α' which itself has spin $\boldsymbol{\Omega}$ against the adopted frame α. On this view, $\boldsymbol{\omega}' \wedge \mathbf{r}$ is the distribution of particle velocity about P seen from α', while $\boldsymbol{\Omega} \wedge \mathbf{r}$ is the velocity in α of the corresponding fixed point of α'.

In applications, when there is some special direction $\mathbf{l}$ fixed in the body, it is often convenient to regard $\boldsymbol{\omega}$ as compounded of a spin about $\mathbf{l}$ together with a spin about the perpendicular direction in the plane of $\boldsymbol{\omega}$ and $\mathbf{l}$. For instance, $\mathbf{l}$ may indicate an axis of geometric symmetry or an actual shaft on which the body turns. Such a decomposition is given by the identity

$$\boldsymbol{\omega} \equiv (\mathbf{l}\boldsymbol{\omega})\mathbf{l} + \mathbf{l} \wedge (\boldsymbol{\omega} \wedge \mathbf{l})$$

which becomes

$$\boldsymbol{\omega} = \nu\mathbf{l} + \mathbf{l} \wedge \dot{\mathbf{l}} \tag{4}$$

when (1) holds. The component spin $\mathbf{l}\boldsymbol{\omega}$ about the special direction is here written as ν; the perpendicular component (its direction generally *not* fixed in the body) evidently has magnitude $|\dot{\mathbf{l}}|$ since $\dot{\mathbf{l}}$ is orthogonal to unit vector **l**. In this latter part of the motion all particles are carried parallel to the tangent plane of the (non-circular) conical surface swept out by **l**. In this sense formula (4) uniquely expresses the total relative motion through the rotation of an embedded line together with a spin about that line. Conversely, one can clearly show this from first principles and then, using the distributive property of vector products, obtain yet another proof of Euler's theorem.

As the motion continues, the spin axis drawn through any considered particle P generates a conical surface. This is called the **space cone** for the motion about P relative to a stated frame; it need not be circular and may not even close (for instance, it can be spiral). At each instant some embedded direction at P is momentarily stationary and coincides with the spin axis; the embedded surface formed in the body by all such directions is called the **body cone** at P, and is also usually neither circular nor closed. The cones not only have a generator in common but actually touch (if their angle of intersection were not zero, the finite spin of the body cone about the intersection could not transport another of its generators on to the space cone in a further arbitrarily small interval of time). We may say, indeed, that the body cone *rolls* on the space cone in the motion relative to P, since the associated velocity of all particles on the line of tangency vanishes momentarily. Each cone consists of a single axis when, and only when, this relative motion is planar, that is, such as to preserve the orientation of one and the same family of parallel planes at all times.

As to the distribution of *absolute* velocity, little need be said. It is fully specified when, as well as the spin, there is given the velocity of one selected particle in the adopted frame. Thus, if $\mathbf{u}(t)$ is the velocity of a reference particle P, the generic particle Q at distance $\mathbf{r}$ from P has velocity

$$\mathbf{v} = \mathbf{u} + \boldsymbol{\omega} \wedge \mathbf{r}, \qquad (5)$$

by superimposing the rotation (1) on the translation $\mathbf{u}$. Note the formal resemblance to the wellknown moment distribution of a system of forces: the linear resultant corresponds to $\boldsymbol{\omega}$ (both being free vectors), the moments about P and Q to $\mathbf{u}$ and $\mathbf{v}$ respectively. For this reason properties in statics have their kinematic analogues. It is enough to mention two. First, the scalar product $\mathbf{v}\boldsymbol{\omega}$ is an invariant, $\mathbf{u\omega}$, in that its value is uniform through the body; this merely gives formal expression to the evident fact that all particles have the same component motion parallel to the spin axis. It follows that the body generally contains no particle at rest, the invariant being unrestricted. Second, since particles whose mutual position vector is in this direction have no relative velocity, the absolute velocity $\mathbf{v}$ has a constant value on each $\boldsymbol{\omega}$-line.

Now imagine the body indefinitely extended, with relation (5) formally continued through all space. We can identify a unique $\boldsymbol{\omega}$-line (perhaps wholly outside the body proper) that contains points whose motion is purely longitudinal, the transverse component of $\mathbf{u}$ exactly annulling the rotation there about P. This line is easily located through the appropriate decomposition

$$\mathbf{u} \equiv [\boldsymbol{\omega}(\mathbf{u}\boldsymbol{\omega}) + \boldsymbol{\omega} \wedge (\mathbf{u} \wedge \boldsymbol{\omega})]/\boldsymbol{\omega}^2, \quad \boldsymbol{\omega} \neq \mathbf{0}.$$

Taking this with (5) it is apparent that the particle velocity is $\boldsymbol{\omega}(\mathbf{u}\boldsymbol{\omega})/\boldsymbol{\omega}^2$ on the $\boldsymbol{\omega}$-line through the point at distance $(\boldsymbol{\omega} \wedge \mathbf{u})/\boldsymbol{\omega}^2$ from P. The analysis shows that a general rigid motion is equivalent to a 'screw' when the invariant is not zero, in analogy to a 'wrench' in statics. The motion is either a pure translation or a pure rotation when and only when the invariant vanishes. This happens, for example, when the body always moves parallel to a fixed plane or turns about a fixed pivot or rolls on a fixed surface. The $\boldsymbol{\omega}$-line coinciding with particles then momentarily at rest is called the **instantaneous axis** of rotation.

Lastly, the generalization of results in §2.1 can now be stated. Let $\boldsymbol{\Omega}(t)$ be the spin of a frame α relative to a frame β. Let $\mathbf{p}$ be the vector representation of some physical entity and denote its rates of change in the two frames by $(\dot{\mathbf{p}})_\alpha$ and $(\dot{\mathbf{p}})_\beta$. Each of these is the vector whose components on cartesian axes fixed in the respective frame are the time derivatives of the components of $\mathbf{p}$. Thus, $(\dot{\mathbf{p}})_\alpha \delta t$ and $(\dot{\mathbf{p}})_\beta \delta t$ are the apparent changes in $\mathbf{p}$ during an infinitesimal time δt, as seen from the two frames. Their difference must accordingly be the change seen from β of a vector $\mathbf{p}$ rotating rigidly with α, namely $(\boldsymbol{\Omega} \wedge \mathbf{p})\delta t$. Whence

$$(\dot{\mathbf{p}})_\beta = (\dot{\mathbf{p}})_\alpha + \boldsymbol{\Omega} \wedge \mathbf{p}, \tag{6}$$

which corresponds with (1) in §2.1. The relationship is of course reciprocal and could equally well be shown as

$$(\dot{\mathbf{p}})_\alpha = (\dot{\mathbf{p}})_\beta + (-\boldsymbol{\Omega}) \wedge \mathbf{p}$$

with $-\boldsymbol{\Omega}$ read as the spin of β relative to α.

In particular, by taking for $\mathbf{p}$ the mutual spin itself,

$$(\dot{\boldsymbol{\Omega}})_\alpha = (\dot{\boldsymbol{\Omega}})_\beta. \tag{7}$$

From the viewpoint of α, regarded as a frame embedded in a rigid body, the left side is parallel to the tangent plane of the current $\boldsymbol{\Omega}$ generator of the body cone. From the viewpoint of β the right side is tangential to the space cone for the motion of α relative to β. The equality expresses that the cones touch, as already remarked.

To obtain the connexion between the second rates of change we apply the above formula to the first rate of change in β,

$$\left(\frac{d}{dt}\right)_\beta (\dot{\mathbf{p}})_\beta = \left\{\left(\frac{d}{dt}\right)_\alpha + \boldsymbol{\Omega} \wedge\right\} (\dot{\mathbf{p}})_\beta$$

and substitute (6) on the right side before differentiating. The result is

$$(\ddot{\mathbf{p}})_\beta = (\ddot{\mathbf{p}})_\alpha + \boldsymbol{\Omega} \wedge (\boldsymbol{\Omega} \wedge \mathbf{p}) + \dot{\boldsymbol{\Omega}} \wedge \mathbf{p} + 2\boldsymbol{\Omega} \wedge (\dot{\mathbf{p}})_\alpha \tag{8}$$

where, by (7), $\dot{\boldsymbol{\Omega}}$ can be written indifferently for the apparent spin-rate in either frame. This is the required generalization of (2) in §2.1. By taking for $\mathbf{p}$ the position vector between two points we recover the connexion between apparent accelerations (§§2.1 and 2.8).

3.2 Dynamical Basis

We return to the Eulerian equations of balance and insert into them the velocity distribution for a rigid body. For convenience the equations of §2.10 are re-stated here:

$$\int \dot{\mathbf{v}}\, dm = \mathbf{R}, \qquad \int \mathbf{r} \wedge \dot{\mathbf{v}}\, dm = \mathbf{M}(P), \tag{1}$$

where $\dot{\mathbf{v}}$ is the acceleration of a generic mass-element dm in position $\mathbf{r}$ with respect to P, the arbitrary origin of moments. $\mathbf{R}$ is the linear resultant, and $\mathbf{M}(P)$ the couple at P, of the actual and fictitious body-forces $\boldsymbol{\gamma}$ per unit mass and the surface forces $\boldsymbol{\tau}$ per unit area:

$$\mathbf{R} = \int \boldsymbol{\gamma}\, dm + \int \boldsymbol{\tau}\, dS, \quad \mathbf{M}(\mathrm{P}) = \int \mathbf{r} \wedge \boldsymbol{\gamma}\, dm + \int \mathbf{r} \wedge \boldsymbol{\tau}\, dS. \tag{2}$$

According to §3.1 the velocity is expressible as

$$\mathbf{v} = \mathbf{u}(P) + \boldsymbol{\omega} \wedge \mathbf{r} \tag{3}$$

for any choice of P, where $\boldsymbol{\omega}$ is the spin and $\mathbf{u}(P)$ is the velocity that the body has at P, or would have if it extended as far.

We introduce also a class of distributions

$$\mathbf{v}^* = \mathbf{u}^*(P) + \boldsymbol{\omega}^* \wedge \mathbf{r} \tag{4}$$

where $\mathbf{u}^*(P)$ and $\boldsymbol{\omega}^*$ are arbitrary constant vectors. For the sake of a mental image these are assigned the dimensions of velocity and spin respectively, so that $\mathbf{v}^*$ can be conceived as being the most general **virtual velocity** admissible in a rigid body (without regard to imposed constraints). The corresponding **virtual work-rate** of the given forces is defined to be

$$\int \boldsymbol{\gamma}\mathbf{v}^* dm + \int \boldsymbol{\tau}\mathbf{v}^* dS.$$

In this we substitute (4) and cyclically permute the terms in the scalar triple products to exhibit the vector products present in (2). After transferring the constant vectors outside the separate integrals there results

$$\mathbf{R}\mathbf{u}^*(P) + \boldsymbol{\omega}^*\mathbf{M}(P). \tag{5}$$

As it stands, this is the virtual work-rate of the force and couple resultants at P, in association with the local velocity and the spin. Thus, in evaluating the total work-rate, the body and surface forces can be replaced by any statically equivalent distribution, and in particular by their resultants at any convenient station. For this purpose it is also immaterial at what point in its line of action a force is considered to be applied, since all particles in that line have the same component of virtual velocity along it (their mutual distances being unvarying).

The virtual work-rate of the mass-accelerations is analogously defined to be

$$\int \dot{\mathbf{v}}\mathbf{v}^* dm.$$

By once again using (4) and interchanging factors in the triple product, this becomes

$$\left(\int \dot{\mathbf{v}} dm\right)\mathbf{u}^*(P) + \boldsymbol{\omega}^*\left(\int \mathbf{r} \wedge \dot{\mathbf{v}} dm\right). \tag{6}$$

Having regard to (1) we see from (5) and (6) that

$$\int \dot{\mathbf{v}}\mathbf{v}^* dm = \int \boldsymbol{\gamma}\mathbf{v}^* dm + \int \boldsymbol{\tau}\mathbf{v}^* dS, \tag{7}$$

which is known as the **principle of virtual work-rate** for a single rigid body. Conversely, if (7) were given for arbitrary distributions (4), the equations of balance would follow as a consequence. For, in (5) and (6), the components of $\mathbf{u}^*$ and $\boldsymbol{\omega}^*$ are independently variable and their respective coefficients are accordingly equal.

Since the class (4) necessarily contains the actual motion at each instant, the actual work-rate of the forces is calculable as

$$\mathbf{R}\mathbf{u}(P) + \boldsymbol{\omega}\mathbf{M}(P)$$

for any P, by specializing (5). By the principle just established this is equal to the actual work-rate of the mass-accelerations,

$$\int \dot{\mathbf{v}}\mathbf{v}dm = \frac{dT}{dt}$$

where T as usual denotes kinetic energy. Thus, for a rigid body, the rate of change of kinetic energy is precisely equal to the rate of working of all body forces and surface forces. This result was previously remarked as a corollary to a theorem for a deformable continuum (§2.12); the present derivation makes no reference to the internal state of stress.

Next we show how the actual value of the kinetic energy is obtainable, without an *ad hoc* computation, once the moment of momentum is known about some point P. By (3)

$$\int \mathbf{v}^2 dm = \int \mathbf{v}(\mathbf{u} + \boldsymbol{\omega} \wedge \mathbf{r})dm = \mathbf{u}\int \mathbf{v}dm + \boldsymbol{\omega}\int \mathbf{r} \wedge \mathbf{v}dm.$$

The first right-hand integral is the linear momentum while the second is the moment of momentum about P, by (2) and (4) in §2.9, remembering that the present $\mathbf{r}$ is measured from P to a generic element. Therefore

$$2T = m\mathbf{v}_G\mathbf{u}(P) + \boldsymbol{\omega}\mathbf{h}(P). \tag{8}$$

Similarly, for the motion $\dot{\mathbf{r}} = \boldsymbol{\omega} \wedge \mathbf{r}$ *relative* to the particle at P, the energy and moment of momentum are such that

$$\left.\begin{aligned} 2T_{\text{rel}}(P) = \int \dot{\mathbf{r}}^2 dm, \quad \mathbf{h}_{\text{rel}}(P) = \int \mathbf{r} \wedge \dot{\mathbf{r}}dm, \\ 2T_{\text{rel}}(P) = \boldsymbol{\omega}\mathbf{h}_{\text{rel}}(P). \end{aligned}\right\} \tag{9}$$

In particular, when P is at G, (9) gives the rotational energy and (8) is König's decomposition, (6) in §2.12. Again, when the body contains a particle O momentarily at rest, so that the kinematic invariant vanishes, (8) and (9) applied to O coincide in

$$2T = \boldsymbol{\omega}\mathbf{h}(O) \quad (\mathbf{v} = \mathbf{0} \text{ at } O). \tag{10}$$

The moment of relative momentum depends linearly on the spin. To emphasize this the following notation is adopted:

$$\left.\begin{aligned} \mathbf{h}_{\text{rel}}(P) &= \int \mathbf{r} \wedge (\boldsymbol{\omega} \wedge \mathbf{r}) dm \equiv I(P)\boldsymbol{\omega}, \\ 2T_{\text{rel}}(P) &= \boldsymbol{\omega} I(P) \boldsymbol{\omega}, \end{aligned}\right\} \tag{11}$$

where I is a 3×3 matrix and $\boldsymbol{\omega}$ is regarded as a row or column, as appropriate. In particular, when there is a particle O momentarily at rest, the *absolute* energy and moment of momentum are

$$\mathbf{h}(O) = I(O)\boldsymbol{\omega}, \qquad 2T = \boldsymbol{\omega} I(O) \boldsymbol{\omega}. \tag{12}$$

The moment of absolute momentum about a general point follows from (7) in §2.9:

$$\mathbf{h}(P) = I(P)\boldsymbol{\omega} - \boldsymbol{\rho}_P \wedge m\mathbf{u}(P) \tag{13}$$

where $\mathbf{u}$ is the body velocity at P and the position vector $\boldsymbol{\rho}_P$ is from G towards P. Alternatively, from (6) in §2.9,

$$\mathbf{h}(P) = I(G)\boldsymbol{\omega} - \boldsymbol{\rho}_P \wedge m\mathbf{v}_G. \tag{14}$$

The coefficients of I will be obtained in §3.4; for the present it is enough to note that they depend on the orientation of the body to the coordinate axes and on the mass distribution about the considered point. When I is not merely a scalar multiple of the unit matrix (and it only is when P is at G and there happens to be all-round kinetic symmetry there), the moment of relative momentum is not usually parallel to the spin. [For instance, when the axis of $\boldsymbol{\omega}$ is fixed in the adopted frame, and hence also in the body turning round it, the direction of $I\boldsymbol{\omega}$ at any P is itself embedded at a fixed angle to $\boldsymbol{\omega}$ and so generates a circular cone in space as the rotation continues.] The relation between the spin and the moment of relative momentum is always one-to-one, in that a unique inverse of matrix I exists; for, since $2T_{\text{rel}} = \boldsymbol{\omega} I \boldsymbol{\omega}$ is intrinsically positive by the definition (9), $I\boldsymbol{\omega}$ cannot vanish for any $\boldsymbol{\omega} \neq \mathbf{0}$.

With the same notation the rotational equation of balance is

$$\frac{d}{dt}\{I(G)\boldsymbol{\omega}\} - \boldsymbol{\rho}_P \wedge m\dot{\mathbf{v}}_G = \mathbf{M}(P) \tag{15}$$

from (4) in §2.10. It is recalled that this is valid in all circumstances, the position of the origin of moments being arbitrary and its motion irrelevant. When a point O of the body is *permanently* at rest in the adopted frame of reference a simpler variant is available:

$$\frac{d}{dt}\{I(O)\boldsymbol{\omega}\} = \mathbf{M}(O) \quad (\mathbf{v} \equiv \mathbf{0} \text{ at } O) \tag{16}$$

since the moment of mass-acceleration about such an origin is equal to the rate of change of the moment of momentum. Again, when there is need to display the properties of a body in relation to some special *particle P*, the appropriate variant is

$$\frac{d}{dt}\{I(P)\boldsymbol{\omega}\} - \boldsymbol{\rho}_P \wedge m\dot{\mathbf{u}}(P) = \mathbf{M}(P) \tag{17}$$

obtained by combining (11) above with (12) in §2.9.

All these formulae call for the evaluation of $d\mathbf{h}_{\text{rel}}/dt$ with respect to some particle of the body. This derivative can be reduced further by going back to the definition (9):

$$\frac{d\mathbf{h}_{\text{rel}}}{dt} = \int \mathbf{r} \wedge \ddot{\mathbf{r}}\,dm = \int \mathbf{r} \wedge \frac{d}{dt}(\boldsymbol{\omega} \wedge \mathbf{r})dm$$

$$= \int \mathbf{r} \wedge (\dot{\boldsymbol{\omega}} \wedge \mathbf{r})dm + \boldsymbol{\omega} \wedge \int (\mathbf{r} \wedge \dot{\mathbf{r}})dm.$$

In the last triple product the first two factors have been interchanged, as is permissible since both are perpendicular to the third. The leading integral is recognizably the same function of the spin-acceleration as $\mathbf{h}_{\text{rel}}$ in (11) is of the spin, while the final integral is just $\mathbf{h}_{\text{rel}}$ itself. Accordingly, the equation can be put as

$$\frac{d}{dt}(I\boldsymbol{\omega}) = I\dot{\boldsymbol{\omega}} + \boldsymbol{\omega} \wedge (I\boldsymbol{\omega}). \tag{18}$$

The terms on the right represent the contributory rates of change due respectively to variations in spin and in orientation relative to the adopted frame. The second term, $\boldsymbol{\omega} \wedge (I\boldsymbol{\omega})$, can indeed be none other than an explicit expression for $(dI/dt)\boldsymbol{\omega}$, where dI/dt is the matrix of time-derivatives of the varying components of I on fixed coordinates in the adopted frame; details of this equivalence are shown later (§3.6 and Exercise 10). The vector product can also be seen as a direct 'convection effect' wherein the vector $I\boldsymbol{\omega}$, considered to be embedded in the body, is rotated about the spin axis. Correspondingly, according to the standpoint of (6) in §3.1, the term $I\dot{\boldsymbol{\omega}}$ must be the apparent rate of change of $I\boldsymbol{\omega}$ in a frame spinning with the body. And this is readily verified since matrix I has unvarying components on embedded axes, *ipso facto*, while the apparent spin-acceleration in the body frame is identical with that in the adopted frame, by applying (7) in §3.1 to the body spin.

3.3 Application of the Principles

Before pursuing generalities it is useful to see the basic principles applied in a few simple though non-trivial situations. For this purpose, avoiding further analysis of the matrix I, the body is taken to be a sphere with centro-symmetric distribution of density. It is supposed subject to the earth's field and to be in point contact with either a fixed or moving rigid constraint, sufficiently rough to prevent slipping. The frame of reference is terrestrial but Coriolis effects of the earth's rotation are ignored [see (15) in §2.9]. Such problems have intrinsic dynamical interest and some practical relevance. Moreover, their solution demands certain kinematical techniques of wide applicability.

With the stipulated symmetry the geometric and mass centres coincide, while the moment of momentum about G is coaxial with the spin whatever its direction in the body. The factor of proportionality is the familiar moment of inertia about a diameter, best computed as $\frac{2}{3}$ of the polar moment of inertia about G (cf. §1.16). We write

$$\mathbf{h}(G) = mk^2\boldsymbol{\omega} \tag{1}$$

where $$mk^2 = \tfrac{2}{3}\textstyle\int r^2 dm,$$

k being the 'radius of gyration' and r the central distance. For example, with a as the radius of the sphere, $(k/a)^2$ is $\frac{2}{5}$ when the density is uniform ($dm \propto r^2 dr$) and $\frac{2}{3}$ when all the mass is concentrated in a surface layer. The matrix $I(G)$ is here diagonal for any choice of cartesian coordinates and just $mk^2 \times$ the unit matrix.

The equations of balance, (3) in §2.10, reduce in the terrestrial frame to

$$\dot{\mathbf{v}}_G = \mathbf{g} + \mathbf{c}, \qquad k^2\dot{\boldsymbol{\omega}} = \mathbf{c} \wedge a\mathbf{n}, \tag{2}$$

where $m\mathbf{g}$ is the weight, $m\mathbf{c}$ the unknown reaction at the point of contact P, and $\mathbf{n}$ the unit normal at P towards G (Fig. 19). If the reaction is of no interest, a single equation of balance suffices:

$$k^2\dot{\boldsymbol{\omega}} + a\mathbf{n} \wedge \dot{\mathbf{v}}_G = a\mathbf{n} \wedge \mathbf{g} \tag{3}$$

by elimination. This is otherwise obtained directly by comparing moments about P, as in (4) in §2.10 or (15) in §3.2; the convective term vanishes identically, the orientation of the sphere being obviously without significance.

Where surfaces touch, three velocities must as a rule be distinguished: $\mathbf{v}_P$ for the *geometric point P* marking the location, and

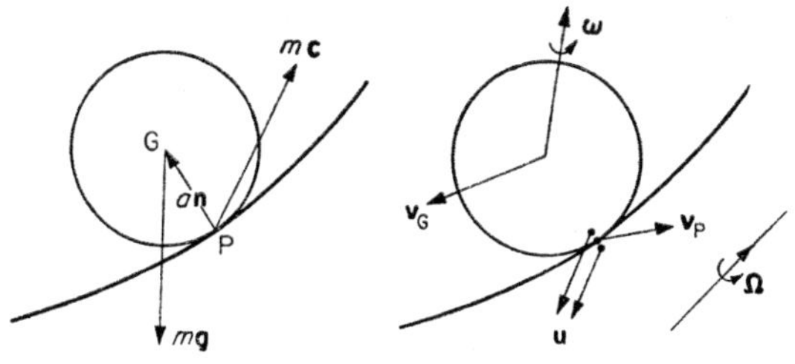

FIG. 19

$\mathbf{u}_1$, $\mathbf{u}_2$ for the respective *material particles* momentarily coinciding with P. In rolling contact, $\mathbf{u}_1$ and $\mathbf{u}_2$ are the same, say $\mathbf{u}$ (though the particle accelerations differ); the equality itself can be formulated in *any* convenient frame of reference, not necessarily the adopted one, since both particles occupy the same position. When, as here, one surface is an inexorable constraint, $\mathbf{u}$ is a pre-ordained function of position and time. The velocity of any other particle in the sphere is related to it by (5) in §3.1, which for the mass-centre is

$$\mathbf{v}_G - \mathbf{u} = \boldsymbol{\omega} \wedge a\mathbf{n}. \tag{4}$$

Equations (3) and (4) are the basis for the solution.

If required, the motion of point P is obtained from

$$\mathbf{v}_G - \mathbf{v}_P = a\dot{\mathbf{n}}, \tag{5}$$

which simply expresses the rate of change of the directed segment $PG = a\mathbf{n}$ as the difference of its terminal velocities. By subtracting (5) from (4) the rate at which P traces a locus on the sphere is found as

$$\mathbf{v}_P - \mathbf{u} = -a(\dot{\mathbf{n}} - \boldsymbol{\omega} \wedge \mathbf{n}). \tag{6}$$

This is otherwise seen without reference to G from the rate of change of $\mathbf{n}$ relatively to the material of the sphere, by (6) in §3.1.† Similarly, when the surface of constraint is itself spherical, with prescribed spin $\mathbf{\Omega}(t)$ and radius b, its centre being on the same side,

$$\mathbf{v}_P - \mathbf{u} = -b(\dot{\mathbf{n}} - \mathbf{\Omega} \wedge \mathbf{n}). \tag{7}$$

Whence

$$\left.\begin{aligned} (a-b)\dot{\mathbf{n}} &= (a\boldsymbol{\omega} - b\mathbf{\Omega}) \wedge \mathbf{n}, \\ \left(\frac{1}{a} - \frac{1}{b}\right)(\mathbf{v}_P - \mathbf{u}) &= (\boldsymbol{\omega} - \mathbf{\Omega}) \wedge \mathbf{n}. \end{aligned}\right\} \tag{8}$$

The first of these is needed when (4) is differentiated for insertion in (3). The second is an analogue of an elementary formula in plane

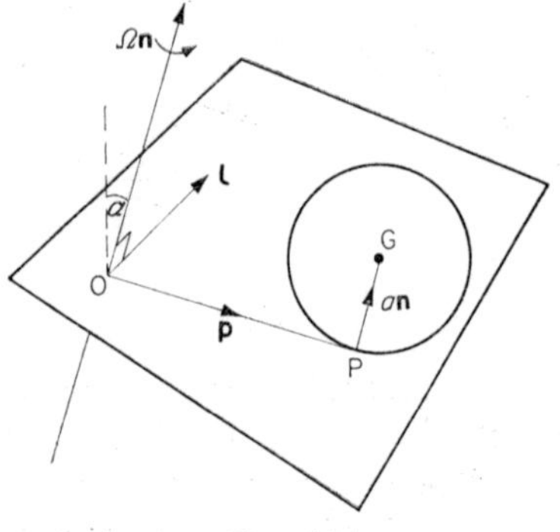

FIG. 20

motion, usually presented as the speed of advance of an instantaneous centre along the centrodes. Notice that the component spins about the common normal naturally do not enter either equation, since they can be varied independently without affecting the rolling contact.

(i) *Rotated plane constraint*

Suppose the constraint to be a plane with constant inclination α to the horizontal and rotated at uniform rate Ω about a normal axis through a fixed point O (Fig. 20). The orbit of G is then necessarily

† The result is in fact an *intrinsic* property of a surface. In general the relative rate of change of the unit normal has components $-(\kappa_1 v_1, \kappa_2 v_2)$ on the orthogonal lines of curvature, where (v_1, v_2) are the corresponding components of relative velocity of the moving point and (κ_1, κ_2) are the principal curvatures (both $1/a$ for a sphere).

in the parallel plane at distance a. Let $\mathbf{p}$ denote the vector perpendicular of G and P from the axis of rotation. The particles in contact have the pre-ordained velocity $\mathbf{u} = \Omega\mathbf{n} \wedge \mathbf{p}$, and the first of (8) is replaced by $\dot{\mathbf{n}} = \mathbf{0}$ since the normal has constant direction in the earth's frame. From (4)

$$\mathbf{v}_G = \mathbf{n} \wedge (\Omega\mathbf{p} - a\boldsymbol{\omega}), \qquad \dot{\mathbf{v}}_G = \mathbf{n} \wedge (\Omega\mathbf{v}_G - a\dot{\boldsymbol{\omega}}), \tag{9}$$

since $\dot{\mathbf{p}} = \mathbf{v}_G (= \mathbf{v}_P)$. Observe that the time-derivative of $\mathbf{u}$ following the path of P is *not* the centripetal acceleration of the particle of the plane momentarily at P.

To obtain the orbit of G we eliminate the body spin between (3) and the second of (9), remembering that $\mathbf{n}\dot{\mathbf{v}}_G = 0$:

$$\dot{\mathbf{v}}_G = \frac{k^2\Omega}{a^2 + k^2}\,\mathbf{n} \wedge \left(\mathbf{v}_G - \frac{a^2 g \sin\alpha}{k^2\Omega}\,\mathbf{l}\right) \tag{10}$$

where $\mathbf{l}$ is the horizontal unit vector perpendicular to the axis of rotation in the sense shown. Thus, relatively to a point C moving parallel to $\mathbf{l}$ with constant speed $a^2 g \sin\alpha/k^2\Omega$, G describes a circle at uniform angular rate $k^2\Omega/(a^2 + k^2)$ if $\Omega \neq 0$; the radius and initial position of C are determined by the conditions of projection. [To reach this solution most easily, compare (10) with the prototype $\dot{\mathbf{v}} = \boldsymbol{\lambda} \wedge \mathbf{v}$ where $\boldsymbol{\lambda}\mathbf{v} = 0$, $\mathbf{v} = \dot{\mathbf{r}}$ being a velocity and $\boldsymbol{\lambda}$ a vector constant. An integration produces $\mathbf{v} = \boldsymbol{\lambda} \wedge (\mathbf{r} - \mathbf{r}_0)$ which evidently represents a spin $\boldsymbol{\lambda}$ about an arbitrary origin $\mathbf{r}_0$ so that the considered point must describe a circle of arbitrary radius about this axis of rotation.] As for the spin of the sphere itself, (2) implies $\mathbf{n}\dot{\boldsymbol{\omega}} = 0$ and so, since $\mathbf{n}$ is here constant, the spin $\mathbf{n}\boldsymbol{\omega}$ about the normal retains its initial value. The remaining component parallel to the plane varies in a manner obtainable from the first of (9) via the solution of (10).

When, however, the rotated plane is horizontal the orbit of G is a circle fixed in the earth's frame. In particular G can remain at rest, the sphere having constant spin $\Omega\mathbf{p}/a$. Again, when the plane does not rotate but is inclined, the orbit is a parabola whose axis is down a line of greatest slope. For, in analogy to a particle under gravity, the acceleration (10) is always in this direction and has constant magnitude $a^2 g \sin\alpha/(a^2 + k^2)$.

(ii) *Fixed spherical constraint*

Suppose that the sphere rolls inside a fixed spherical container with radius b and centre O. Then $\mathbf{\Omega}$ and $\mathbf{u}$ are both zero, while

$$\mathbf{v}_G = (a - b)\dot{\mathbf{n}} = a\boldsymbol{\omega} \wedge \mathbf{n}. \tag{11}$$

We notice an immediate first integral:

$$\boldsymbol{\omega}\mathbf{n} = \text{constant}, \quad \nu \text{ say}, \tag{12}$$

since $\dot{\boldsymbol{\omega}}\mathbf{n} = 0$ from (2) and $\boldsymbol{\omega}\dot{\mathbf{n}} = 0$ from (11). (This constancy of the normal component of spin is not found, however, when the container is non-spherical.)

Another first integral is supplied by the work principle in the earth's frame. Since the work-rate of the contact reaction vanishes with the local particle velocity, the kinetic energy T and potential energy $m(b - a)\mathbf{ng}$ with respect to O have a constant sum. This is of course not additional to (3) but derivable from it, through scalar multiplication by the spin and use of (11). Now

$$\frac{2T}{m} = \mathbf{v}_G^2 + k^2\boldsymbol{\omega}^2 = \left(1 + \frac{k^2}{a^2}\right) \mathbf{v}_G^2 + k^2\nu^2$$

and so

$$\frac{1}{2}\left(1 + \frac{k^2}{a^2}\right)\mathbf{v}_G^2 + (b - a)\mathbf{ng} = \text{constant}, \tag{13}$$

on transferring the unvarying part of T to the right-hand side. A third integral follows from the observation that, since $\dot{\mathbf{n}} \wedge \mathbf{v}_G = \mathbf{0}$ by (11), the equation of balance (3) can be put as

$$\frac{d}{dt}\mathbf{h}(P) = a\mathbf{n} \wedge m\mathbf{g} \tag{14}$$

where

$$\mathbf{h}(P)/m = k^2\boldsymbol{\omega} + a\mathbf{n} \wedge \mathbf{v}_G$$

$$= k^2\nu\mathbf{n} + \left(1 + \frac{k^2}{a^2}\right) a\mathbf{n} \wedge \mathbf{v}_G$$

by (11) and (12). Actually, this is just the variant (9) in §2.9 since G and the point P move parallel to one another, being always collinear with O. By resolving (14) on the vertical we see that $\mathbf{gh}(P)$ does not vary, or that

$$a\left(1+\frac{k^2}{a^2}\right)(\mathbf{g}\wedge\mathbf{n})\mathbf{v}_G + k^2\nu\mathbf{ng} = \text{constant}, \tag{15}$$

after permuting the scalar triple product. The component moment of momentum about the vertical through O is, by contrast, not constant since the frictional component of the contact force does not generally lie in the meridian plane and so has a moment about this vertical.

Equations (13) and (15) completely define the orbit of G on the sphere of radius $b - a$ and centre O. They resemble the corresponding integrals for a spherical pendulum (§2.5), exactly so when $\nu = 0$, and can be manipulated in a similar way when examining the path stability and range. For this the meridional and transverse components of velocity of G are introduced, together with the inclination θ of OG to the downward vertical; then $\mathbf{ng} = -g\cos\theta$ and the triple product $= g\sin\theta \times$ the transverse velocity. However, further discussion is left until §3.9, when the precisely analogous equations of motion of a symmetrical top are treated.

The overriding requirement that the reaction should have a positive normal component must be kept in mind. The analysis is given in §2.5 but may be noted again in the present context. We are concerned with the sign of

$$\mathbf{nc} = \mathbf{n}(\dot{\mathbf{v}}_G - \mathbf{g})$$

from (1). Since G always moves perpendicularly to the radius, both $\mathbf{nv}_G$ and its derivative vanish. The centripetal acceleration is therefore

$$\mathbf{n}\dot{\mathbf{v}}_G = -\dot{\mathbf{n}}\mathbf{v}_G = \mathbf{v}_G^2/(b-a).$$

The required condition is thus

$$\mathbf{v}_G^2 \geqslant (b-a)\mathbf{ng}, \tag{16}$$

which can be further reduced with the help of the energy integral. Properties of the frictional component of the reaction are given in Exercise 8.

3.4 Properties of the Inertia Tensor

The moment of momentum and the kinetic energy of a rigid body, in its motion relative to any of its particles P, have been reduced to the forms

$$\mathbf{h}_{\text{rel}}(P) = I(P)\boldsymbol{\omega}, \qquad 2T_{\text{rel}}(P) = \boldsymbol{\omega} I(P)\boldsymbol{\omega},$$

where $\boldsymbol{\omega}$ is the body spin (§3.2). $I(P)$ is the symmetric matrix defined by either of the equivalances

$$I(P)\mathbf{l} \equiv \int \mathbf{r} \wedge (\mathbf{l} \wedge \mathbf{r})dm, \qquad \mathbf{l}I(P)\mathbf{l} \equiv \int (\mathbf{l} \wedge \mathbf{r})^2 dm, \tag{1}$$

for any (unit) vector $\mathbf{l}$, where $\mathbf{r}$ is the position of the mass element dm with respect to P. The definition has been re-phrased in this way to free it from appearing to be restricted to rigid bodies. In fact the purely geometric properties of the matrix can be studied just as well for a deforming continuum, though their dynamical relevance is naturally less specific.

Let any rectangular axes at P be chosen and the components of $\mathbf{r}$ and $\mathbf{l}$ written as x_i and l_i $(i = 1, 2, 3)$. Then, with the customary summation convention and the Kronecker delta (see §2.11),

$$\left.\begin{aligned} \mathbf{r} \wedge (\mathbf{l} \wedge \mathbf{r}) &= \mathbf{l}\mathbf{r}^2 - (\mathbf{l}\mathbf{r})\mathbf{r} \equiv (r^2\delta_{ij} - x_i x_j)l_j, \\ (\mathbf{l} \wedge \mathbf{r})^2 &= \mathbf{l}^2\mathbf{r}^2 - (\mathbf{l}\mathbf{r})^2 \equiv (r^2\delta_{ij} - x_i x_j)l_i l_j, \end{aligned}\right\} \tag{2}$$

since the scalar product $\mathbf{l}\mathbf{r}$ can be written as $l_i x_i$ or $l_j x_j$. The typical element in the ith row and jth column of the inertia matrix is thus

$$I_{ij}(P) = \int (r^2\delta_{ij} - x_i x_j)dm. \tag{3}$$

The diagonal elements $(i = j)$ were named **moments of inertia** by Euler (c.1741); the others $(i \neq j)$ are sometimes known as moments of deviation, or more usually their negatives are called **products of inertia.** In applications where subscripts can be dispensed with, the notation

$$I \equiv \begin{pmatrix} A & -H & -G \\ -H & B & -F \\ -G & -F & C \end{pmatrix}$$

is customary.† The typical moment and product of inertia are then

$$A \equiv I_{11} = \int(x_2^2 + x_3^2)dm,$$

$$F \equiv -I_{23} = -I_{32} = \int x_2 x_3 dm.$$

The axial moments of inertia have an invariant sum, independent of the choice of coordinates, equal to twice the polar moment of inertia about P:

$$A + B + C = 2J \quad \text{where } J = \int \mathbf{r}^2 dm.$$

Obvious inequalities, typically

$$B + C \geqslant A \geqslant 2|F|,$$

are useful on occasion; the equalities hold only when the mass is distributed in a plane, the first when this is $x_1 = 0$ and the second when it is either $x_2 + x_3 = 0$ or $x_2 - x_3 = 0$.

The immediate task is to find how the matrix elements depend on the orientation of the coordinates within the body. Let any other rectangular axes x_p^* ($p = 1, 2, 3$) be taken at P and denote by l_{pi} the cosine of the angle between the pth new axis and the ith old one. Then l_{pi} ($i = 1, 2, 3$) are the old coordinates of a unit vector along the pth new axis, while l_{qj} ($q = 1, 2, 3$) are the new coordinates of a unit vector along the jth old axis. The magnitudes of such vectors, together with their orthogonality within each triad, are accordingly expressible as

$$l_{pi}l_{qi} = \delta_{pq}, \qquad l_{qi}l_{qj} = \delta_{ij},$$

by forming mutual scalar products. Also, by projection, the linear transformation between the two sets of coordinates is

$$x_p^* = l_{pi}x_i \quad \text{with inverse} \quad x_j = l_{qj}x_q^*,$$

which can be seen as scalar products of $\mathbf{r}$ with the typical new and old unit vectors. There are precisely similar connexions between the sets of components of any vector quantity. However, our concern is with the elements of the inertia matrix for the new coordinates:

$$I_{pq}^*(P) = \int(r^2\delta_{pq} - x_p^* x_q^*)dm.$$

Now $$(x_p^* x_q^*) = (l_{pi}x_i)(l_{qj}x_j) = l_{pi}l_{qj}(x_i x_j)$$

and $$\delta_{pq} = l_{pi}l_{qi} = l_{pi}l_{qj}\delta_{ij},$$

† The symbol G, of course, is also used for the mass-centre, but no confusion need arise.

both of which relations exhibit the transformation rule characterizing a cartesian tensor of second rank [cf. (7) in §2.11]. It follows immediately that the sets of moments of inertia and deviation similarly satisfy

$$I^*_{pq} = l_{pi}l_{qj}I_{ij}, \qquad I_{ij} = l_{pi}l_{qj}I^*_{pq},$$

and themselves constitute a second rank tensor. To emphasize this property under transformation we shall henceforth refer to matrix I as the **inertia tensor.**

It makes for brevity to introduce the orthogonal 3×3 matrix L which has l_{pi} in its pth row and ith column. The columns of L are therefore the new components of the old unit vectors, while the columns of its transpose L' are the old components of the new unit vectors. All preceding relations obtainable by varying the free subscripts can now be brought together as matrix equations:

$$I^* = LIL', \qquad I = L'I^*L, \tag{4}$$

where

$$\mathbf{x}^* = L\mathbf{x}, \qquad \mathbf{x} = L'\mathbf{x}^*, \tag{5}$$

with

$$LL' = U = L'L.$$

Here U is the unit tensor with components δ_{ij} in any coordinates while $\mathbf{x}$ and $\mathbf{x}^*$ are 3×1 column matrices formed by the respective sets of components of vector $\mathbf{r}$. Alternatively to the first of (4) the transformation rule can be shown as

$$I^*_{pq} = \mathbf{l}_p I \mathbf{l}_q \tag{6}$$

where $\mathbf{l}_p$ and $\mathbf{l}_q$ are unit vectors along the pth and qth new axes, or, to be precise, respectively the row and column matrices of their old components. Elements of the second triple product can be expressed in reciprocal fashion. On putting $p = q$ in (6) and suppressing this subscript, the moment of inertia about an axis in direction $\mathbf{l}$ is found as

$$\mathbf{l}I\mathbf{l} \equiv Al_1^2 + Bl_2^2 + Cl_3^2 - 2Fl_2l_3 - 2Gl_3l_1 - 2Hl_1l_2. \tag{7}$$

For clarity the expanded form is given on the right, in terms of components of the inertia tensor and unit vector on (the same) arbitrary coordinates. With a similarly abbreviated notation the moment of deviation in respect of any pair of orthogonal directions $\mathbf{l}$ and $\mathbf{n}$ is

$$\mathbf{l}I\mathbf{n} = \mathbf{n}I\mathbf{l} \equiv Al_1n_1 + \ldots - F(l_2n_3 + l_3n_2) - \ldots. \tag{8}$$

A vector or first rank tensor is ordinarily represented geometrically by a directed segment, whose projections on any rectangular axes x_i are the corresponding 'components' v_i of the quantity. However, the behaviour of these components under transformation of co-ordinates could be pictured just as well in terms of a representative plane. A first rank tensor is associated with an invariant linear form

$$v_p^* x_p^* = (v_i l_{pi}) x_p^* = v_i (l_{pi} x_p^*) = v_i x_i$$

or

$$\mathbf{v}^* \mathbf{x}^* = (\mathbf{v} L') \mathbf{x}^* = \mathbf{v}(L' \mathbf{x}^*) = \mathbf{v}\mathbf{x}$$

by (5). Its components therefore transform like the coefficients in the equation of any fixed plane, $\mathbf{vx}$ = constant, perpendicular to the directed segment. Similarly, for any symmetric tensor of second rank, in particular I_{ij}, there is an invariant quadratic form:

$$\mathbf{x}^* I^* \mathbf{x}^* = \mathbf{x}^*(LIL')\mathbf{x}^* = (\mathbf{x}^* L) I (L' \mathbf{x}^*) = \mathbf{x} I \mathbf{x}$$

by (4) and (5). In this case the components transform like the symmetrized coefficients in the equation of a quadric surface (of arbitrary scale),

$$\mathbf{x} I \mathbf{x} \equiv A x_1^2 + \ldots - F(x_2 x_3 + x_3 x_2) - \ldots = md^4 \text{ say,} \tag{9}$$

which in that sense is a representation of the associated symmetric tensor. The quadratic form in respect of the inertia tensor itself is generally positive-definite and the surface an ellipsoid. For, by putting $\mathbf{x} = r\mathbf{l}$ the left-hand side can be identified with $r^2 \times$ the moment of inertia (7) which is intrinsically positive by definition. Exceptionally, when the mass is distributed along a line, as for an idealized rod, the moment of inertia about the line vanishes and the surface is a cylinder open in that direction.

At any point $\mathbf{x}$ on the inertia ellipsoid in respect of a particle P of the continuum the outward normal has the direction of the vector $I(P)\mathbf{x}$. For, if $\mathbf{x} + \delta\mathbf{x}$ is a neighbouring point also satisfying (9), then $(\delta\mathbf{x}) I \mathbf{x} + \mathbf{x}(I \delta\mathbf{x}) = 0$ to first order. But the bilinear forms, or scalar products, here are equal since the tensor is symmetric, and so $(\delta\mathbf{x}) I \mathbf{x} = 0$. $I\mathbf{x}$ is thus orthogonal to any tangential direction $\delta\mathbf{x}$ and therefore normal; it is clearly outward since (9) states its scalar product with the radius to be positive. In relation to a rigid motion this geometrical property has the interpretation that the moment of relative momentum about P,

$$I(P)\boldsymbol{\omega} \equiv (A\omega_1 - H\omega_2 - G\omega_3, \quad B\omega_2 - F\omega_3 - H\omega_1, \quad C\omega_3 - G\omega_1 - F\omega_2), \quad (10)$$

has the direction of the outward normal where the local inertia ellipsoid is pierced by the spin vector $\boldsymbol{\omega}$ from P (Fig. 21). Generally,

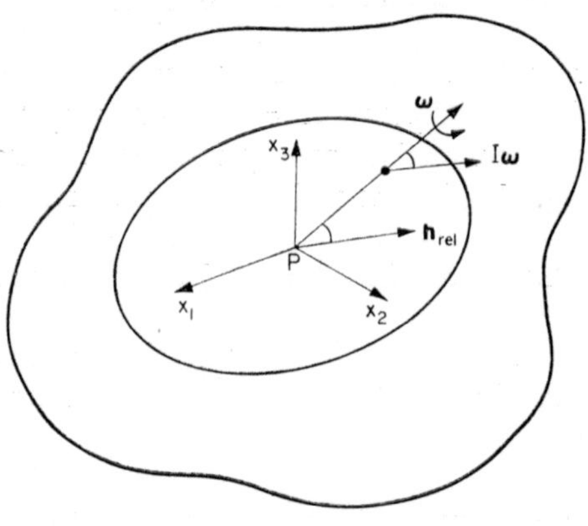

FIG. 21

then, the vector moment of relative momentum deviates from the axis of spin. In fact, its component on any direction $\mathbf{n}$ perpendicular to that axis is $(\mathbf{n}I\mathbf{l})\omega$ where $\boldsymbol{\omega} = \omega\mathbf{l}$. According to (8) the bracketed scalar here is just the 'moment of deviation' associated with this pair of directions, which explains the terminology. For instance, if x_3 is taken along the axis of spin so that $\omega_1 = \omega_2 = 0$, the components of $I\boldsymbol{\omega}$ are $(-G, -F, C)\omega$.

The moment of relative momentum coincides with the spin axis if and only if this lies in a so-called **principal direction** satisfying

$$I\mathbf{l} = mk^2\mathbf{l} \quad (\mathbf{l}^2 = 1) \quad (11)$$

for a scalar k to be determined, evidently the radius of gyration about such an axis. We have in (11) a homogeneous system of 'characteristic' or 'eigenvalue' equations, $(I - mk^2U)\mathbf{l} = \mathbf{0}$; these are treated in texts on linear algebra,† but for convenience the basic theory is recapitulated in the present context. A necessary and sufficient condition that such a set of homogeneous equations has a solution other than zero is that its determinant vanishes. Hence k must be such as to make the 3×3 determinant

$$|I - mk^2U| = 0. \tag{12}$$

Since matrix I is symmetric this cubic equation in k^2 has three real roots, all positive when I is positive-definite; exceptionally, when the mass is distributed along a line, one root is zero and I is only semi-definite and has a vanishing determinant. In any event k itself is real with non-negative values κ_1, κ_2, κ_3, say. If these are distinct, (11) defines a unique direction for each, say $\boldsymbol{\lambda}_1$, $\boldsymbol{\lambda}_2$, $\boldsymbol{\lambda}_3$, such that

$$I\boldsymbol{\lambda}_1 = m\kappa_1^2\boldsymbol{\lambda}_1, \quad I\boldsymbol{\lambda}_2 = m\kappa_2^2\boldsymbol{\lambda}_2, \quad I\boldsymbol{\lambda}_3 = m\kappa_3^2\boldsymbol{\lambda}_3.$$

From the first pair of equations, and since I is symmetric,

$$m\kappa_1^2\boldsymbol{\lambda}_1\boldsymbol{\lambda}_2 = \boldsymbol{\lambda}_2 I\boldsymbol{\lambda}_1 = \boldsymbol{\lambda}_1 I\boldsymbol{\lambda}_2 = m\kappa_2^2\boldsymbol{\lambda}_1\boldsymbol{\lambda}_2,$$

which imply

$$\boldsymbol{\lambda}_1\boldsymbol{\lambda}_2 = 0 = \boldsymbol{\lambda}_1 I\boldsymbol{\lambda}_2,$$

and similarly for the other two pairs. The triad of directions is therefore mutually orthogonal, and all associated moments of deviation vanish as arranged at the outset. Consequently, when this triad is taken for coordinate axes, the matrix of components of the inertia tensor becomes purely diagonal, with $m\kappa_1^2$, $m\kappa_2^2$, $m\kappa_3^2$ as the leading elements or moments of inertia, and the equation of the surface (9) is reduced to a sum of squares. In short, these directions are the geometric axes of the inertia ellipsoid, as was already apparent in (11) which states their normality to the surface. If, on the other hand, (12) has a repeated root, say $\kappa_2 = \kappa_3$, then (11) yields an associated family of directions forming a unique plane normal to $\boldsymbol{\lambda}_1$; the quadric is then a spheroid, whose axis of symmetry and any perpendicular through P are principal directions. When all the roots coincide, the quadric is a sphere. In such cases the body is said to

† For example D. C. Murdoch: *Linear Algebra for Undergraduates* (John Wiley and Sons, Inc., New York 1957) and D. T. Finkbeiner: *Introduction to Matrices and Linear Transformations* (W. H. Freeman and Co., San Francisco and London 1960).

possess axial or spherical kinetic symmetry, respectively, in relation to P.

The next task is to establish connexions between the inertia tensors and principal directions at different points of the continuum. These all stem from the **theorem of parallel axes**, which is another decomposition formula resting on the special property of the mass-centre, (1) in §2.9. To obtain it, substitute $\mathbf{r} = \boldsymbol{\rho} - \boldsymbol{\rho}_P$ in (1), where as usual $\boldsymbol{\rho}$ denotes the generic position vector from G and the subscript refers to the considered particle P. By straightforward reduction

$$\left.\begin{aligned} I(P)\mathbf{l} &= I(G)\mathbf{l} + m\boldsymbol{\rho}_P \wedge (\mathbf{l} \wedge \boldsymbol{\rho}_P), \\ \mathbf{l}I(P)\mathbf{l} &= \mathbf{l}I(G)\mathbf{l} + m(\mathbf{l} \wedge \boldsymbol{\rho}_P)^2. \end{aligned}\right\} \tag{13}$$

Alternatively, and with more insight, these can be seen as structurally like (14) in §3.2 and König's formula, respectively, written for an arbitrary rigid motion relative to P with $\mathbf{l}$ along the axis of spin. From either, having regard to (2), follows

$$I_{ij}(P) = I_{ij}(G) + m(\boldsymbol{\rho}^2\delta_{ij} - \xi_i\xi_j)_P \tag{14}$$

where the coordinate axes at G and P are parallel and ξ_i denotes components of $\boldsymbol{\rho}$. Or this is obtained directly from (3), remembering that the mass integral of each ξ_i vanishes separately. Stating (14) in words: the inertia tensor at P is equal to that at G plus that at P due to a mass-point m at G. In detail: at a generic point the moment of inertia about the x_1 direction exceeds that at G by $m(\xi_2^2 + \xi_3^2)$, while the product of inertia for the x_2 and x_3 directions exceeds that at G by $m\xi_2\xi_3$. Correspondingly, the equation of the inertia quadric at a generic point can be expressed as

$$\mathbf{x}I(G)\mathbf{x} + m(\mathbf{x} \wedge \boldsymbol{\rho})^2 = md^4, \tag{15}$$

by the second of (13) applied to (9). The inertia tensor at any P due to a mass-point m at G clearly has axial symmetry about their join; the principal moments of inertia at P due to m are 0 about GP and $m(GP)^2$ about any perpendicular through P. Consequently, (14) implies that the principal directions at P for the actual body are generally oriented differently from those at G. Only when direction GP is itself principal at G is it also principal at P and both triads parallel.

In illustration consider a uniform solid cube with side $2a$. From the geometric symmetry the directions through G parallel to the edges are principal at G, and the corresponding moments of inertia are all equal, in fact to $\frac{2}{3}ma^2$ by the wellknown elementary formula. The inertia ellipsoid at G is therefore a sphere and all-round kinetic symmetry exists there with *any* direction principal. In particular, the long diagonal to any corner C is principal at G and hence also at the corner itself, where there is axial kinetic symmetry (and indeed at all points of the cube). The distance from G being $\sqrt{3}a$, the parallel axes theorem gives the principal moments of inertia at C as $\frac{2}{3}ma^2$ about the diagonal and $(\frac{2}{3} + 3)ma^2 = 11ma^2/3$ about any perpendicular. If, however, the elements of the inertia tensor there were required for coordinates along the edges, (14) would be applied with $\boldsymbol{\rho} = a(1, 1, 1)$:

$$\frac{I(C)}{ma^2} = \begin{pmatrix} \frac{2}{3} & 0 & 0 \\ 0 & \frac{2}{3} & 0 \\ 0 & 0 & \frac{2}{3} \end{pmatrix} + \begin{pmatrix} 2 & -1 & -1 \\ -1 & 2 & -1 \\ -1 & -1 & 2 \end{pmatrix}.$$

The moments of inertia about the edges are thus $8ma^2/3$ and the products of inertia ma^2 at C. If, now the characteristic determinant (12) were to be evaluated in these coordinates, the principal moments already obtained would be recovered and (11) found to be satisfied by directions such that $l_1 = l_2 = l_3$ or $l_1 + l_2 + l_3 = 0$, corresponding to the long diagonal and the plane perpendicular to it.

3.5 Impulse and Inertia

It is desirable to acquire facility in handling the inertia tensor before proceeding to general problems in continued motion. A context well-suited to this purpose is the dynamics of a rigid body under sudden impact or impulsive constraint. The classical theory is again mainly due to Euler.

The term 'impulse', in its broad meaning, refers to the time integral of a force over an arbitrary interval. We write

$$\mathbf{L}(t, t^0) = \int_{t^0}^{t} \mathbf{F}(t)dt$$

for the impulse of a force $\mathbf{F}$ during the time between some initial instant t^0 and a later instant t. Its dimensions are evidently those of mass × velocity. Indeed, if the force acts only on a mass-point, the impulse is precisely equal to the accompanying change in linear momentum, by integrating once the Second Law. Similarly, for a continuum under point loads, the linear equation of balance gives

$$m\Big[\mathbf{v}_G\Big]_{t^0}^{t} = \Sigma\mathbf{L}. \tag{1}$$

Square brackets here indicate the difference of terminal values, in accordance with the standard usage in integral calculus; henceforward the upper and lower limits will be omitted. The impulses of body forces, such as the weight, are not included in the summation, on the understanding that they are far exceeded during this interval by the applied loads, as usually happens.

On the other hand, the rotational equation of balance with respect to the mass-centre has only the formal integral

$$[\mathbf{h}(G)] = \Sigma\int \boldsymbol{\rho} \wedge \mathbf{F}dt,$$

where $\boldsymbol{\rho}$ is the vector drawn from G to a typical point of application, supposed always coincident with the same particle. However, motivated by the wish to express also these integrals in terms of the impulses, we impose the restrictions that the motion is rigid and that the amount of rotation during the interval is negligible. Then the relative position vectors $\boldsymbol{\rho}$ remain constant in direction as well as magnitude and can be taken outside the integrals:

$$[\mathbf{h}(G)] \simeq \Sigma(\boldsymbol{\rho} \wedge \int\mathbf{F}dt) = \Sigma(\boldsymbol{\rho} \wedge \mathbf{L}). \tag{2}$$

The effect of this approximation is to produce equations sufficient to determine the overall change in motion when only the total impulses are given. The detailed variations in the contributory forces during the interval are no longer of concern; in fact such data are rarely available in practice for actual impacts. Of course, in any application of the theory the validity of neglecting body-forces and angular displacements remains to be confirmed. The scope of the theory is, in fact, quite closely circumscribed; usually, for example, it is at least necessary that the interval should be small compared with a representative period of oscillation of the body as a compound pendulum, but large compared with the time of passage of elastic waves across the body. However, in the conventional theory all appropriate conditions are tacitly assumed to be met.

On multiplying (1) vectorially by $\boldsymbol{\rho}_P$, drawn from G to some other origin of moments P, and subtracting from (2) there results

$$[\mathbf{h}(P)] = \Sigma(\mathbf{r} \wedge \mathbf{L})$$

where $\mathbf{r} = \boldsymbol{\rho} - \boldsymbol{\rho}_P$ is from P to a point of application. (When the translational displacement during the interval is not also small, P can be imagined to move with the body.) In view of the assumed rigidity and constant orientation a more explicit form is

$$I(G)[\boldsymbol{\omega}] - m\boldsymbol{\rho}_P \wedge [\mathbf{v}_G] = \Sigma(\mathbf{r} \wedge \mathbf{L}), \tag{3}$$

since the inertia tensor at G has the same components on fixed axes throughout the interval. A frequently useful variant is

$$I(P)[\boldsymbol{\omega}] - m\boldsymbol{\rho}_P \wedge [\mathbf{u}] = \Sigma(\mathbf{r} \wedge \mathbf{L}) \tag{4}$$

from (13) in §3.2, where $\mathbf{u}$ is the velocity of the material particle at P. Or this follows from (3) above and (13) in §3.4, since

$$\mathbf{u} = \mathbf{v}_G + \boldsymbol{\omega} \wedge \boldsymbol{\rho}_P.$$

Again, if there happens to be a particular particle O which is permanently fixed, the impulsive moment about it is equal simply to $I(O)[\boldsymbol{\omega}]$.

Even though the overall displacement may be infinitesimal the applied forces nevertheless do a finite amount of work. In fact, the total work could be put formally as

$$\Sigma\int_{t^0}^{t} \mathbf{F}\mathbf{v}\,dt = \Sigma(\mathbf{L}\bar{\mathbf{v}})$$

where $\mathbf{v}$ is the varying velocity of the particle at a point of application and $\bar{\mathbf{v}}$ is some appropriate average over the interval. However, since the variations in time of both force and velocity are neither known nor even small, the integral cannot be reduced directly. But, whatever the intervening motion might be, it is naturally subject to the general principle that the total work is equal to the jump in energy (here entirely kinetic by hypothesis). Now the translational and rotational parts of this are

$$[T_{\text{trans}}] = \tfrac{1}{2}[m\mathbf{v}_G^2] = \tfrac{1}{2}(\mathbf{v}_G^0 + \mathbf{v}_G)m[\mathbf{v}_G],$$

$$[T_{\text{rot}}] = \tfrac{1}{2}[\boldsymbol{\omega}I(G)\boldsymbol{\omega}] = \tfrac{1}{2}(\boldsymbol{\omega}^0 + \boldsymbol{\omega})I(G)[\boldsymbol{\omega}]$$

since $\boldsymbol{\omega}^0 I \boldsymbol{\omega} = \boldsymbol{\omega} I \boldsymbol{\omega}^0$ by the tensor symmetry, where the superscript distinguishes an initial value. Whence, by (1) and (2),

$$\left.\begin{aligned}[T] &= \tfrac{1}{2}(\mathbf{v}_G^0 + \mathbf{v}_G)\Sigma\mathbf{L} + \tfrac{1}{2}(\boldsymbol{\omega}^0 + \boldsymbol{\omega})\Sigma(\boldsymbol{\rho} \wedge \mathbf{L}) \\ &= \Sigma\{\tfrac{1}{2}(\mathbf{v}^0 + \mathbf{v})\mathbf{L}\}\end{aligned}\right\} \tag{5}$$

since $\mathbf{v} = \mathbf{v}_G + \boldsymbol{\omega} \wedge \boldsymbol{\rho}$ at any time. Thus, the required average turns out to be the simple mean of the terminal velocities. The overall increase in energy can be calculated, according to convenience, from either right-hand expression; their equivalence is in exact analogy with (5) in §3.2 for a system of forces.

It will be noticed that the classical theory of impulse reduces every problem to a system of linear algebraic equations. To take the simplest example: suppose a body which is free to turn about a fixed pivot is set in motion from rest by an impulsive moment $\mathbf{h}$ about O. Then, if I denotes the inertia tensor at O, the spin when the impulse ceases is to be found from $I\boldsymbol{\omega} = \mathbf{h}$; the instantaneous axis is therefore conjugate to the plane of the impulsive moment (Fig. 21). Specifically, if J denotes the tensor inverse of I such that

$$IJ = JI = U \tag{6}$$

the spin is $\boldsymbol{\omega} = J\mathbf{h}$.† Together with the energy T generated by the impulse the entire solution can be recapitulated as

$$I\boldsymbol{\omega} = \mathbf{h}, \qquad \boldsymbol{\omega} = J\mathbf{h}, \qquad 2T = \boldsymbol{\omega}\mathbf{h}. \tag{7}$$

The duality here suggests the introduction of a quadric surface associated with J:

$$\mathbf{y}J\mathbf{y} = \frac{1}{m}, \tag{8}$$

with respect to which the impulsive moment and plane of rotation are conjugate. This quadric is called the **ellipsoid of gyration** and is the polar reciprocal of the ellipsoid of inertia, (9) in §3.4. The vector radii to corresponding points defined by

$$I\mathbf{x} = md^2\mathbf{y}, \qquad J\mathbf{y} = \mathbf{x}/md^2, \tag{9}$$

have an invariant scalar product

$$\mathbf{x}\mathbf{y} = d^2,$$

† The symbol J is used in Chapter I for a polar moment of inertia: this does not appear in the present context.

since this can be shown as $\mathbf{x}I\mathbf{x}/md^4$ or $md^2\mathbf{y}J\mathbf{y}$. In geometrical terms, either radius is orthogonal to the tangent plane through the terminus of the other and meets both planes in inverse points with respect to the sphere of radius d. From this follows the characteristic property of the ellipsoid of gyration: the radius of gyration about any direction $\mathbf{x}$ is equal to the central perpendicular on the tangent plane at $\mathbf{y}$. In proof observe that, by (7) in §3.4, the moment of inertia is $\mathbf{x}I\mathbf{x}/|\mathbf{x}|^2 = md^4/|\mathbf{x}|^2$, while the length of the perpendicular is $d^2/|\mathbf{x}|$ since its terminus is inverse to the point $\mathbf{x}$.

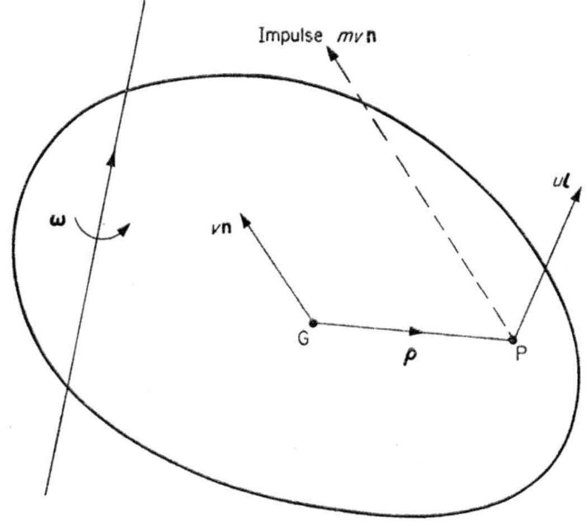

FIG. 22

(i) *Free motion initiated by impulse*

A blow is applied at the point P of an arbitrary rigid body initially at rest and unconstrained (Fig. 22). Let the final velocity of P be written as $u\mathbf{l}$ and the impulse as $mv\mathbf{n}$. Here m is the mass, u and v have dimensions of velocity, and $\mathbf{l}$ and $\mathbf{n}$ are unit vectors. Then by (1) $v\mathbf{n}$ is the eventual velocity of the mass-centre G.

Suppose, in the first place, that the impulse is prescribed. Then by moments about G the spin is determined as

$$I(G)\boldsymbol{\omega} = mv\boldsymbol{\rho} \wedge \mathbf{n} \quad \text{or} \quad \boldsymbol{\omega} = mvJ(G)(\boldsymbol{\rho} \wedge \mathbf{n}) \tag{10}$$

where $\boldsymbol{\rho}$ is the position vector GP. The velocity produced at P follows from kinematics as $v\mathbf{n} + \boldsymbol{\omega} \wedge \boldsymbol{\rho}$, or

$$u\mathbf{l} = v\{\mathbf{n} - \boldsymbol{\rho} \wedge mJ(G)(\boldsymbol{\rho} \wedge \mathbf{n})\}, \tag{11}$$

which generally deviates from the line of action of the blow. The energy delivered is

$$\tfrac{1}{2}mv^2 + \tfrac{1}{2}\boldsymbol{\omega} I(G)\boldsymbol{\omega} = \tfrac{1}{2}mv^2\{1 + (\boldsymbol{\rho} \wedge \mathbf{n})mJ(G)(\boldsymbol{\rho} \wedge \mathbf{n})\}, \tag{12}$$

either directly from König's formula or the first of (5).

Suppose, alternatively, that the final velocity of P is prescribed instead of the impulse. The spin is now determined most readily from (4) by moments about P:

$$I(P)\boldsymbol{\omega} = mu\boldsymbol{\rho} \wedge \mathbf{l} \quad \text{or} \quad \boldsymbol{\omega} = muJ(P)(\boldsymbol{\rho} \wedge \mathbf{l}). \tag{13}$$

The appropriate impulse $mv\mathbf{n}$ is obtained via the velocity of G, which is $v\mathbf{n} = u\mathbf{l} - \boldsymbol{\omega} \wedge \boldsymbol{\rho}$ or

$$v\mathbf{n} = u\{\mathbf{l} + \boldsymbol{\rho} \wedge mJ(P)(\boldsymbol{\rho} \wedge \mathbf{l})\}. \tag{14}$$

This is, of course, ultimately equivalent to (11). A corresponding expression for the energy, from the second of (5), is $\tfrac{1}{2}muv\mathbf{ln}$ or

$$\tfrac{1}{2}mu^2 - \tfrac{1}{2}\boldsymbol{\omega} I(P)\boldsymbol{\omega} = \tfrac{1}{2}mu^2\{1 - (\boldsymbol{\rho} \wedge \mathbf{l})mJ(P)(\boldsymbol{\rho} \wedge \mathbf{l})\}. \tag{15}$$

It is of interest to consider the extremal values of the energy when $\mathbf{n}$ is varied in (12) but v is held fixed, or when $\mathbf{l}$ is varied in (15) but u is held fixed. Since the quadratic forms are non-negative, the minimum of (12) or the maximum of (15) is attained when the respective form vanishes. This requires the spin to be zero and the motion therefore a pure translation, with the blow directed along GP in either sense. On the other hand, for a maximum of (12) or a minimum of (15), $\mathbf{n}$ or $\mathbf{l}$ respectively must plainly lie in the plane through G or P perpendicular to their join (whatever the direction of each vector product its magnitude is greatest when the factors are orthogonal, since their amplitudes are fixed). Further, since the perpendicular sections of the ellipsoids of gyration at G and P are ellipses, each extremum is attained when the vector product is directed along the respective minor axis, which requires $\mathbf{n}$ or $\mathbf{l}$ (as the case may be) to be along the respective major axis. Analytically, the axes of section are defined by

$$\mathbf{n}J(G)(\mathbf{n} \wedge \boldsymbol{\rho}) = 0 = \mathbf{n}\boldsymbol{\rho}, \quad \text{or} \quad \mathbf{l}J(P)(\mathbf{l} \wedge \boldsymbol{\rho}) = 0 = \mathbf{l}\boldsymbol{\rho}, \tag{16}$$

since orthogonal diameters are principal if they are also conjugate. Equivalently, recalling (8) in §3.4, each bilinear expression may be read as the tensor component associated with the diameters concerned. In geometrical terms, the required axes are located as the meeting of the perpendicular plane with a certain quadric cone. Finally, on combining (16) with (10) and (13) in turn, we see that each spin axis is at right angles to the path of G or P respectively. The kinematic invariant accordingly vanishes and an instantaneous axis of rotation exists.

(ii) *Motion modified by impulsive constraint*

A body is initially free and in arbitrary motion specified by the spin $\boldsymbol{\omega}^0$, together with the velocity $\mathbf{u}^0$ of a certain particle P in position $\boldsymbol{\rho}$ with respect to the mass-centre G. P is brought suddenly to rest by a suitable impulse applied there. The spin $\boldsymbol{\omega}$ in the subsequent instantaneous rotation is to be determined.

By equating moments about P, the variant (4) being most convenient,

$$I(P)(\boldsymbol{\omega} - \boldsymbol{\omega}^0) = -m\boldsymbol{\rho} \wedge \mathbf{u}^0. \tag{17}$$

Alternatively, the conservation of moment of absolute momentum about P could be shown explicitly as

$$I(P)\boldsymbol{\omega} = I(G)\boldsymbol{\omega}^0 - m\boldsymbol{\rho} \wedge \mathbf{v}^0$$

with the initial and final values on the right and left, respectively, where $\mathbf{v}^0 = \mathbf{u}^0 - \boldsymbol{\omega}^0 \wedge \boldsymbol{\rho}$ is the initial velocity of G. The equivalence with (17) is readily checked from the fundamental identity (13) in §3.4. When (17) has been solved for the final spin, the required impulse is obtained from the jump in linear momentum,

$$\mathbf{L} = -m\{\mathbf{u}^0 + (\boldsymbol{\omega} - \boldsymbol{\omega}^0) \wedge \boldsymbol{\rho}\},$$

since the velocity of G changes from $\mathbf{v}^0$ to $\boldsymbol{\rho} \wedge \boldsymbol{\omega}$. The reduction in energy is afterwards best computed from the second of (5) as $-\frac{1}{2}\mathbf{L}\mathbf{u}^0$; it is also expressible as

$$\tfrac{1}{2}m\{(\mathbf{u}^0)^2 - (\boldsymbol{\rho} \wedge \mathbf{u}^0)mJ(P)(\boldsymbol{\rho} \wedge \mathbf{u}^0)\}$$

analogously to (15). It will be observed that neither the impulse nor the change in energy depend on the initial spin but only on the velocity of the particle P.

The algebra simplifies when GP is an axis of kinetic symmetry. Let k be the radius of gyration about it and κ the radius of gyration about any transverse axis through G. Then the principal moments of inertia at P may be formally decomposed as

$$m(\kappa^2 + \rho^2)\{(1, 1, 1) + (0, 0, \lambda)\} \quad \text{where } 1 + \lambda = \frac{k^2}{\kappa^2 + \rho^2}.$$

From this standpoint the transformation

$$I(P)[\boldsymbol{\omega}] = m(\kappa^2 + \rho^2)[\boldsymbol{\omega} + \lambda(\mathbf{l}\boldsymbol{\omega})\mathbf{l}]$$

becomes self-evident, where $\mathbf{l}$ is a unit vector along GP. The part $(1 + \lambda)(\mathbf{l}\boldsymbol{\omega})\mathbf{l}$ of the quantity in square brackets clearly arises from the component moment of momentum about GP. And this is conserved since the impulse has no moment about this axis; the same also follows by resolving (17). Accordingly, the axial component of spin is unaffected and so

$$I(P)[\boldsymbol{\omega}] = m(\kappa^2 + \rho^2)[\boldsymbol{\omega}]$$

simply. Returning to (17) we now have at once

$$\boldsymbol{\omega} = \boldsymbol{\omega}^0 - (\boldsymbol{\rho} \wedge \mathbf{u}^0)/(\kappa^2 + \rho^2).$$

The necessary impulse at P is

$$\mathbf{L} = -\frac{m\kappa^2}{\kappa^2 + \rho^2}\left\{\mathbf{u}^0 + \frac{\rho^2}{\kappa^2}(\mathbf{l}\mathbf{u}^0)\mathbf{l}\right\},$$

which has its line of action through the point $-\kappa^2\mathbf{u}^0/(\boldsymbol{\rho}\mathbf{u}^0)$ with respect to G.

(iii) *Collision with a fixed constraint*

A plane surface of constraint is frictionless and fixed. A body of mass m travels directly towards the plane with speed u^0 and strikes it at a point P. The impulse can be written as $mv\mathbf{n}$, where $\mathbf{n}$ is the unit outward normal, while the final velocity $\mathbf{u}$ is assumed to be tangential. The unknowns v, $\mathbf{u}$ and the spin $\boldsymbol{\omega}$ are to be determined.

By the linear equation of balance the final velocity of the mass-centre G is $(v - u^0)\mathbf{n}$. Whence, by kinematics,

$$\mathbf{u} = (v - u^0)\mathbf{n} + \boldsymbol{\omega} \wedge \boldsymbol{\rho}, \qquad \mathbf{n}\mathbf{u} = 0, \tag{18}$$

where $\boldsymbol{\rho}$ is the vector GP, and so

$$v + \{\mathbf{n}\boldsymbol{\omega}\boldsymbol{\rho}\} = u^0. \tag{19}$$

By moments about G,

$$I(G)\boldsymbol{\omega} = mv\boldsymbol{\rho} \wedge \mathbf{n}. \tag{20}$$

In order to eliminate the spin between the last two equations we form again the triple product appearing in (19):

$$\boldsymbol{\omega} I(G)\boldsymbol{\omega} = mv\{\mathbf{n}\boldsymbol{\omega}\boldsymbol{\rho}\} = mv(u^0 - v).$$

Returning once more to (20):

$$\boldsymbol{\omega} I(G)\boldsymbol{\omega} = mv^2(\boldsymbol{\rho} \wedge \mathbf{n})mJ(G)(\boldsymbol{\rho} \wedge \mathbf{n}).$$

Equating the last two expressions:

$$v = u^0/\{1 + (\boldsymbol{\rho} \wedge \mathbf{n})mJ(G)(\boldsymbol{\rho} \wedge \mathbf{n})\}. \tag{21}$$

Thus v is generally less than u^0, so that G still has a residual motion towards the plane. Exceptionally, when G is initially situated on the normal at P, the whole body is brought to rest. Otherwise the induced spin is given by (20) with (21), and lastly the slipping velocity by (18).

3.6 Differential Equations of Euler

We return to the rotational equation of balance for a general rigid body in continuous motion [§3.2, equation (15) or (17), supplemented by (18)]. The integration of this vector equation is usually to be simplified by resolving it on cartesian axes. To retain fixed coordinate directions throughout the motion would be disadvantageous, however: the orientation of the body changes and with it the relevant moments and products of inertia, whose analytic expressions are in consequence awkward. Instead, it was proposed by Euler that the equation of balance in the adopted frame should be resolved on the current directions of the principal axes of inertia at some definite particle of the body.

The corresponding components of the local inertia tensor I are then unvarying, the products of inertia being zero and the moments of inertia equal to the principal values A, B, C, say. Also, the components of the moment of relative momentum $I\boldsymbol{\omega}$ are just $(A\omega_1, B\omega_2, C\omega_3)$, where $(\omega_1, \omega_2, \omega_3)$ now denote the principal axes components of the spin relative to the adopted frame. The convective part of $d(I\boldsymbol{\omega})/dt$ is thus

$$\boldsymbol{\omega} \wedge I\boldsymbol{\omega} \equiv \{(C - B)\omega_2\omega_3, (A - C)\omega_3\omega_1, (B - A)\omega_1\omega_2\}. \tag{1}$$

As remarked before, these are basically the components of $(dI/dt)\boldsymbol{\omega}$, and in fact the rate of change of I relative to the adopted frame is

$$\frac{dI}{dt} \equiv \begin{pmatrix} 0 & (A-B)\omega_3 & (C-A)\omega_2 \\ (A-B)\omega_3 & 0 & (B-C)\omega_1 \\ (C-A)\omega_2 & (B-C)\omega_1 & 0 \end{pmatrix}$$

resolved on the principal axes. This can be shown directly without difficulty† or by specializing the formula obtained in Exercise 10. The resolution of the remaining part of the rate of change of $I\boldsymbol{\omega}$ is simply

$$I\frac{d\boldsymbol{\omega}}{dt} \equiv (A\dot{\omega}_1, \quad B\dot{\omega}_2, \quad C\dot{\omega}_3)$$

where $(\dot{\omega}_1, \dot{\omega}_2, \dot{\omega}_3)$ are the straightforward time-derivatives of the components of spin. These form a vector representing the rate of change of $\boldsymbol{\omega}$ relative to either the embedded coordinates or the adopted frame (§3.1). In all, therefore, we have $A\dot{\omega}_1 + (C - B)\omega_2\omega_3$ as the first component of $d(I\boldsymbol{\omega})/dt$ and similar expressions for the other components. These can be used in any of the variants (15), (16), or (17) in §3.2.

When the origin of moments is either the mass-centre G or a particle O permanently at rest,

$$\left.\begin{aligned} A\dot{\omega}_1 + (C - B)\omega_2\omega_3 &= M_1, \\ B\dot{\omega}_2 + (A - C)\omega_3\omega_1 &= M_2, \\ C\dot{\omega}_3 + (B - A)\omega_1\omega_2 &= M_3, \end{aligned}\right\} \qquad (2)$$

when the coordinates are the principal axes at G or O respectively. (M_1, M_2, M_3) are the component moments of all forces associated with the frame in which the body has spin $\boldsymbol{\omega}$. As these equations stand, the moving axes are being used merely in an auxiliary role,

† A vector $\mathbf{a}$ fixed in the adopted frame changes at rate $-\boldsymbol{\omega} \wedge \mathbf{a}$ relative to the body. Use this in the time-derivative of $Al_1n_1 + Bl_2n_2 + Cl_3n_3$ and afterwards put $\mathbf{l}$ and $\mathbf{n}$ equal to (1, 0, 0), (0, 1, 0) or (0, 0, 1).

not as a new frame of reference. An observer consistently taking the standpoint of the embedded frame would see the body always at rest, and for his equation of balance would re-write (2) with all left-hand terms transferred to the right and re-interpreted as additional fictitious body-moments *vis-à-vis* the original frame. The character of the various spin terms clearly indicates that on this view $-A\dot{\omega}_1$, etc., are the moment resultants of Euler body-forces while $(B - C)\omega_2\omega_3$, etc., are resultants of centrifugal body-forces (cf. Exercise 7). No Coriolis contribution appears since there is no motion relative to the embedded frame.

The non-linear convective terms are responsible for most of the analytic difficulties of integrating Euler's equations. But the character of the moment resultant is also critical. In this respect a broad distinction is worth making at the outset. Suppose that the moment components are prescribed as functions solely of the time and the spin components on the *embedded* axes, entirely irrespective of the orientation of the body in the adopted frame. Then the equations as they stand are of first order only in the components of spin. According to a basic theorem for systems of ordinary differential equations, a solution exists (and is unique for given initial spin) in some time interval where the moments are piecewise-continuous functions of all the variables and have piecewise-continuous first partial derivatives with respect to the components of spin. The solution determines the body cone and the manner in which the terminus of the spin vector traces a locus on it. The actual orientation of the body in the adopted frame at a generic instant can be found subsequently, if required, by integration of relations between the calculated spin components and any convenient geometric variables and their own time-derivatives. Suppose, on the other hand, that the moment components on the embedded axes depend at least partly on orientation. Then the geometric variables must be introduced from the start and the spin eliminated in their favour. The resulting equations are now of second order and an analogous existence and uniqueness theorem is available for initial data specifying both spin and orientation.

The orientation of the principal axes, or of any embedded triad, is conveniently specified with respect to an adopted frame α by means of angles also introduced by Euler. In Fig. 23 z is a chosen direction fixed in α while 3 moves with the body. The plane $z3$ is located by an angle ϕ measured right-handedly round z from some reference plane fixed in α. The normal n to plane $z3$, which is the intersection of planes respectively perpendicular to z and 3, is called the line of nodes (as in orbital theory). The axis 3 is specified by ϕ and a further angle θ measured towards it from z. Note carefully that θ is always measured in the same sense round the line of nodes, and so changes sign whenever 3 passes through z. Lastly, the axes 1 and

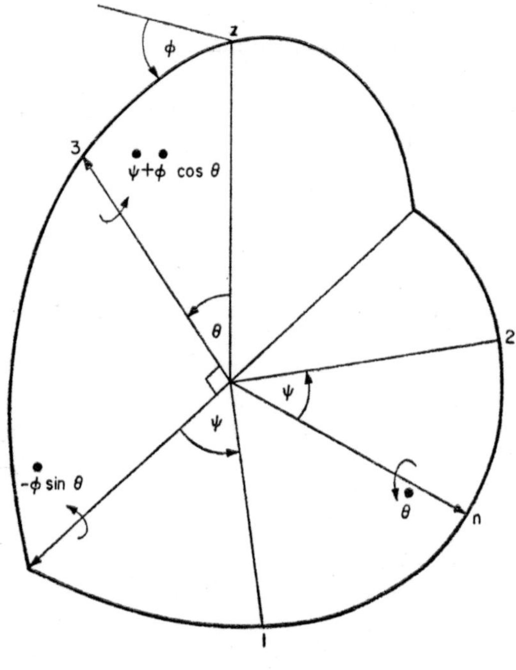

FIG. 23

2 are located by the angle ψ measured right-handedly about 3 from planes $3z$ and $3n$ respectively. (Various notations and sign conventions are in use; in particular, the azimuth is often taken through an additional $\frac{1}{2}\pi$ to the nodal plane zn.) The rotation of the body in α is thereby specified as the vector resultant of spins $\dot{\theta}$ about n, $\dot{\phi}$ about z, and $\dot{\psi}$ about 3 (§3.1), or by the orthogonal components

$$(-\dot{\phi}\sin\theta,\quad \dot{\theta},\quad \dot{\psi}+\dot{\phi}\cos\theta) \tag{3}$$

on the right-handed triad shown in the Figure. The (123) components are therefore

$$\left.\begin{aligned} \omega_1 &= \dot{\theta} \sin \psi - \dot{\phi} \sin \theta \cos \psi, \\ \omega_2 &= \dot{\theta} \cos \psi + \dot{\phi} \sin \theta \sin \psi, \\ \omega_3 &= \dot{\psi} + \dot{\phi} \cos \theta, \end{aligned}\right\} \tag{4}$$

by a further resolution.

Some preliminary illustrations of Euler's method follow.

(i) *Self-excited rotation*

This term relates to any situation where the applied moment about the mass-centre has a fixed direction relative to the body, as in a rocket or space vehicle when the thrust has an embedded line of action. For simplicity we assume that its magnitude is also constant, so that the moment components on the principal axes at G are themselves constant. The frame of reference is sidereal and any gravitational torque is neglected.

We notice first a possible steady-state rotation in which the applied moment does no work and the spin satisfies $\boldsymbol{\omega} \wedge I\boldsymbol{\omega} = \mathbf{M}$. The axis of rotation must therefore have a direction such that

$$\mathbf{M}\boldsymbol{\omega} = 0 = \mathbf{M}I\boldsymbol{\omega},$$

namely normal to the plane of vectors $\mathbf{M}$ and $I\mathbf{M}$ when these are distinct (in sharp contrast with the relation between spin and *impulsive* moment). Any such embedded axis necessarily has fixed orientation also in the adopted frame since it always contains the same particles and they are at rest relative to G. The sense of the spin is arbitrary and its magnitude is found from (1) as the root of

$$(C - B)(B - A)(A - C)\omega^2 = (C - B)^2 \frac{M_2 M_3}{M_1} + (B - A)^2 \frac{M_3 M_1}{M_2} + (A - C)^2 \frac{M_1 M_2}{M_3}$$

provided the sense of $\mathbf{M}$ is such that the root is real. But when $\mathbf{M}$ coincides with a principal direction, say C, the axis of rotation can have any orientation in the transverse plane, with $\omega_1\omega_2 = M/(B - A)$. Indeed, when there is an axis of kinetic symmetry, the $\mathbf{M}$ direction *must* be orthogonal to it and necessarily principal.

The stability of a permanent axis of rotation is investigated in the following manner. Denote the spin in the considered steady state by $\omega\mathbf{l}$ and in a neighbouring motion by $\omega(\mathbf{l}+\boldsymbol{\epsilon})$, where $\boldsymbol{\epsilon}(t)$ is to be treated as a first-order infinitesimal. Omitting terms of higher order in (2),

$$A\dot{\varepsilon}_1 + \omega(C-B)(l_2\varepsilon_3 + l_3\varepsilon_2) = 0$$

with two similar equations. The general solution is an arbitrary superposition of three particular integrals of type

$$\boldsymbol{\epsilon} = \boldsymbol{\eta} e^{\lambda\omega t} \text{ with } \lambda, \boldsymbol{\eta} \text{ constant.}$$

Substitution and cancellation of the time factor leaves linear homogeneous equations in $\boldsymbol{\eta}$, whose determinant must therefore vanish:

$$ABC\lambda^3 + \{A(A-B)(A-C)l_1^2 + \ldots + \ldots\}\lambda$$
$$= 2(A-B)(B-C)(C-A)l_1l_2l_3.$$

Generally at least one root λ has a positive real part, since the second-degree term is absent. This indicates instability, to the extent that the deviation of the spin direction in the motion following an arbitrary infinitesimal disturbance does not remain vanishingly small (cf. §2.5). When the moment is applied about a principal direction, the constant term disappears from the cubic and one root is zero (corresponding to spin about a neighbouring permanent axis in the transverse plane). When, further, the coefficient of λ is negative one other root is positive and there is again instability. On the other hand, when the coefficient is positive, both other roots are pure imaginaries and the linearized solution remains arbitrarily small. But the behaviour of the exact solution remains uncertain without further analysis.

Some idea of the complexity of the motion in general can be had from a complete solution in the case $A = B$, $\mathbf{M} \equiv (0, 0, M)$. By eliminating the non-linear terms between the first pair of Euler equations, the integral

$$\omega_1^2 + \omega_2^2 = \text{constant}, \quad n^2 \text{ say,}$$

is obtained. This is satisfied by introducing a parametric variable $\chi(t)$ such that

$$\omega_1 = n\cos\chi, \quad \omega_2 = n\sin\chi.$$

By returning with these to either of the original pair,

$$\omega_3 = A\dot{\chi}/(C - A).$$

Whence, from the last Euler equation,

$$\chi = \frac{M(C - A)}{2AC}\, t^2, \qquad \omega_3 = \frac{M}{C}\, t,$$

supposing that the initial spin is $n(1, 0, 0)$, say. As χ varies monotonically with time the terminus of the spin vector spirals round an embedded cylinder of radius n about the moment axis. The equation of the corresponding body cone for the rotation about G is found as

$$\frac{x_3^2}{x_1^2 + x_2^2} = \frac{2MA}{n^2C(C - A)} \tan^{-1}\left(\frac{x_2}{x_1}\right)$$

by putting $\omega_1/\omega_3 = x_1/x_3$ and $\omega_2/\omega_3 = x_2/x_3$.

(ii) *Rotation under moving constraint*

An arbitrary body is smoothly mounted on a fine shaft whose direction is principal at a point O. The line of the shaft is made to turn at constant rate Ω round a fixed perpendicular axis z through O. The principal moments of inertia at O are A, B, C where B is about the shaft. The couple resultants about O are (M_1, M_2, M_3), of which M_2 is assumed to be prescribed while M_1 and M_3 are transmitted via the shaft and are to be determined.

The configuration is schematically represented by Fig. 23, with $\dot{\phi}(t) \equiv \Omega$ and $\psi(t) \equiv 0$, the line of nodes being along the shaft and here embedded in the body. By specializing (3) or (4) the principal axes components of spin are

$$\boldsymbol{\omega} \equiv (-\Omega \sin\theta, \quad \dot{\theta}, \quad \Omega \cos\theta)$$

where θ is the inclination of the C axis to the fixed direction z. Equations (2) now give

$$M_1 = (C - A - B)\Omega\dot{\theta} \cos\theta, \quad M_3 = (A - C - B)\Omega\dot{\theta} \sin\theta, \quad (5)$$

where

$$B\ddot{\theta} + (C - A)\Omega^2 \sin\theta \cos\theta = M_2(\theta, \dot{\theta}, t). \quad (6)$$

When the functional dependence of M_2 is prescribed, the rotation about the shaft is found by integrating (6). Afterwards (5) defines the couples that must be applied to the shaft to maintain its stipulated motion; these are seen to be equivalent to couples $B\Omega\dot{\theta}$ and $(A - C)\Omega\dot{\theta}$ about axes in the A, C plane, the first being perpendicular to z and the second inclined to the first at the varying angle 2θ. When observed from the frame turning round z with the shaft, the body appears only to have a spin $\dot{\theta}$ about axis B, and to be subject to additional centrifugal and Coriolis body-force moments which are the negatives of the preceding Ω^2 and $\Omega\dot{\theta}$ terms respectively.

The considered body may for example be a propeller in a ship or aircraft when course is being altered. In this case the mass-centre is on the shaft and M_2 is the net driving torque. When the propeller has three or more symmetrically disposed blades, the shaft is necessarily an axis of kinetic symmetry ($A = C$). Another example is a rotated compound pendulum for which z is the upward vertical and $M_2 = mgd \sin \theta$ supposing G to be situated on the axis C at distance d from O. There is now a first integral:

$$\tfrac{1}{2}B\dot{\theta}^2 + \Psi(\theta) = \text{constant},$$

where
$$\Psi(\theta) = \{mgd - \tfrac{1}{2}(C - A)\Omega^2 \cos \theta\} \cos \theta. \tag{7}$$

This is actually the expression of the work principle in the rotated frame, where the kinetic energy is simply $\tfrac{1}{2}B\dot{\theta}^2$ and the gravitational and centrifugal body-forces have a total potential Ψ. Possible configurations of equilibrium in this frame, equivalent to steady motion in the adopted frame, are found by setting zero $\ddot{\theta}$ in (6) or $\partial\Psi/\partial\theta$ in (7). In addition to the trivial solutions $\theta = 0, \pi$ there is also an inclined position $\theta = \alpha$ when the shaft rotates quickly enough, no couples M_1 and M_3 being needed:

$$\cos \alpha = \frac{mgd}{(C - A)\Omega^2} \quad \text{when} \quad \Omega^2 > \frac{mgd}{|C - A|}.$$

Stability of these configurations can be tested exactly as in §2.5.

3.7 Free Rotation of a General Body

It is supposed here that all the acting forces, real or fictitious, are reducible to a linear resultant at the mass-centre. This is most closely realized when the frame of reference is sidereal, provided that gravitational torques due to other matter are negligible and the body is either in free orbit or so mounted that frictional dissipation is minimal.

In the absence of an applied couple $\mathbf{M}(G)$ the equation of rotational balance states simply that the vector moment of momentum $\mathbf{h} = I(G)\boldsymbol{\omega}$ is conserved. Its axis has therefore an invariable direction in the adopted frame and its magnitude a constant value h, both determined by the initial data. The motion could, for example, be generated from rest by an impulsive couple equal to $\mathbf{h}$. Subsequently the components (h_1, h_2, h_3) on the principal axes at G obviously depend solely on the inclinations of these to the invariable line; so, therefore, do the spin components also, namely

$$(\omega_1, \omega_2, \omega_3) = (h_1/A, h_2/B, h_3/C).$$

The current orientation of the body thereby determines the spin vector, while together they determine a neighbouring orientation after an infinitesimal time, and so on. In this way it can be readily understood in principle how the initial data and invariance of $\mathbf{h}$ govern the ensuing motion.

Since the acting forces have no moment resultant at G they do no net work in the rigid motion relative to it (§3.2). Therefore, by the work principle, the rotational kinetic energy $\frac{1}{2}\boldsymbol{\omega}\mathbf{h}$ is constant, T say (§2.12: the present notation is allowable since the additional and possibly varying translational energy is of no interest). This property could alternatively be deduced from the kinematic identities†

$$\frac{d}{dt}(\tfrac{1}{2}\boldsymbol{\omega}\mathbf{h}) = \boldsymbol{\omega}\frac{d\mathbf{h}}{dt} = \mathbf{h}\frac{d\boldsymbol{\omega}}{dt}$$

together with $d\mathbf{h}/dt = \mathbf{0}$. The terminus of the spin vector is accordingly confined to a fixed plane perpendicular to the invariable line at a distance from G equal to $2T/h$, the constant value of the component spin $(\mathbf{h}/h)\boldsymbol{\omega}$ about that direction (Lagrange). Further, the plane of $\boldsymbol{\omega}$ and $\mathbf{h}$ rotates *monotonically* and right-handedly about $\mathbf{h}$ since $\dot{\boldsymbol{\omega}}$ is always to the side to which $\mathbf{h} \wedge \boldsymbol{\omega}$ points; this can, for instance, be inferred from the sign of the scalar product $\dot{\boldsymbol{\omega}}(\mathbf{h} \wedge \boldsymbol{\omega})$ which, by (18) in §3.2, is equal to the positive quantity $\dot{\boldsymbol{\omega}}I\dot{\boldsymbol{\omega}}$.

But the point $\boldsymbol{\omega}$ also lies on the embedded surface $\boldsymbol{\omega}I\boldsymbol{\omega} = 2T$, which differs only in scale from the ellipsoid of inertia at G. Since its local normal is parallel to the invariable line (as in Fig. 21) the

† The first is already implicit in §3.2. To prove it *ab initio* note that

$$\dot{\mathbf{r}}\ddot{\mathbf{r}} = (\boldsymbol{\omega} \wedge \mathbf{r})\ddot{\mathbf{r}} = \boldsymbol{\omega}(r \wedge \ddot{\mathbf{r}})$$

analogously to the procedure in respect to (6) there.

energy ellipsoid touches the fixed plane, and moreover can be said to roll on it since the point of contact is on the instantaneous axis. The rate of rotation is given by the distance from G to the point of contact, which is decided by the orientation. This pictorial representation, due to Poinsot (1834), gives precise expression to the foregoing qualitative remarks.

The respective loci traced by the point of contact on the energy ellipsoid and on the fixed plane were named by Poinsot the **polhode**

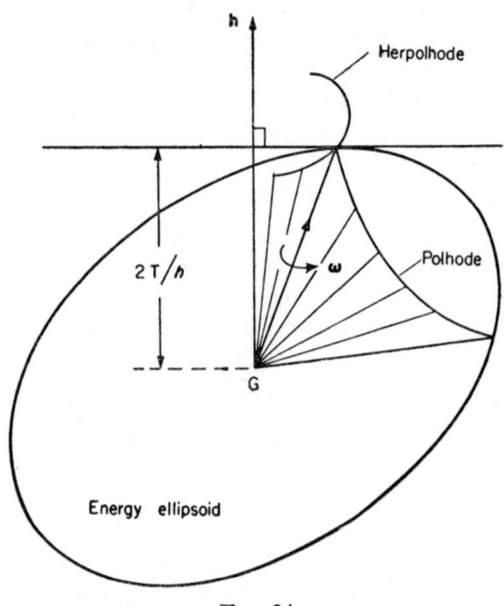

FIG. 24

and **herpolhode** (Fig. 24). They are equally the curves described by the tip of the spin vector on the body and space cones at G (§3.1). The body cone is here a quadric, whose equation is obtained from the conservation integrals

$$A\omega_1^2 + B\omega_2^2 + C\omega_3^2 = 2T, \tag{1}$$

$$A^2\omega_1^2 + B^2\omega_2^2 + C^2\omega_3^2 = h^2, \tag{2}$$

by combining them to form a homogeneous expression:

$$A\left(A - \frac{h^2}{2T}\right)\omega_1^2 + B\left(B - \frac{h^2}{2T}\right)\omega_2^2 + C\left(C - \frac{h^2}{2T}\right)\omega_3^2 = 0. \tag{3}$$

The space cone, however, is not a quadric, nor can its equation be put in simple terms. The invariable line also sweeps out a quadric cone relative to the body, its equation being derived from (3) merely by replacing ω_1 by h_1/A, etc.

The polhode may also be characterized as the intersection of the ellipsoids (1) and (2), or as a curve on (1) such that the central perpendicular on the tangent plane has the same value $2T/h$ at each point. The polhode is clearly a *closed* curve, described monotonically, and the motion of the body is therefore periodic. On the

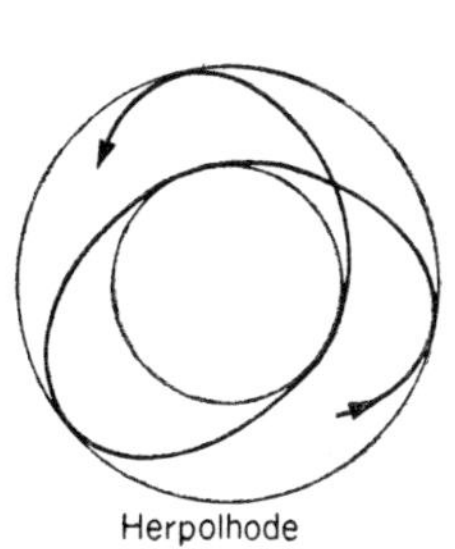

Herpolhode

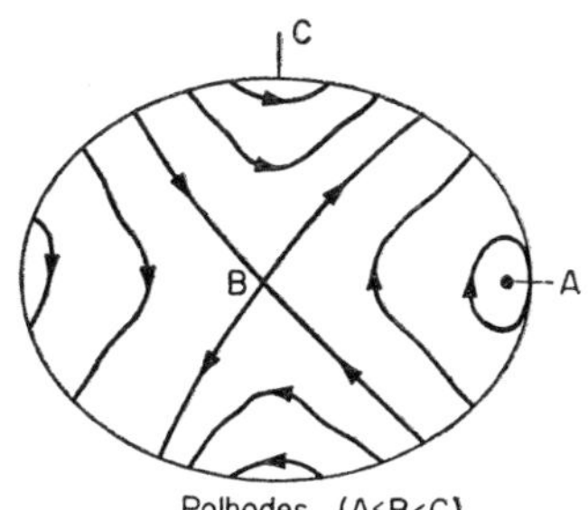

Polhodes $(A<B<C)$

FIG. 25

other hand the herpolhode generally does *not* close. It is a sequence of identical arcs alternately touching two circles which are centred on the invariable line (Fig. 25) and correspond to the extremal radii from G to the polhode. The herpolhode cannot, of course, intersect the invariable line unless this is trivially the permanent axis of rotation, when both loci degenerate to a single point.

Typical polhodes on the energy ellipsoid are shown schematically in Fig. 25 for various values of $h^2/2T$, necessarily between A and C if B is supposedly the intermediate moment of inertia. The polhodes encircle the principal directions A or C according as $h^2/2T$ is in the interval (A, B) or (B, C), and are described right-handedly about the axis of greatest moment, left-handedly about the axis of least moment. They become single points when $h^2/2T = A$ or C since the motion can only be a permanent spin about one or other axis. By contrast when $h^2/2T = B$ a rotation about this axis is not the only possibility: if the initial spin has a different direction the body cone is a pair of planes

$$A(B - A)\omega_1^2 = C(C - B)\omega_3^2 \tag{5}$$

on which the polhodes are segments of ellipses. The motion is no longer periodic and the herpolhode will be shown to be a bounded double spiral, asymptotic to the invariable line ($\omega \to h/B = 2T/h$ when ω_1 and $\omega_3 \to 0$). The sense of description of any polhode, with the ordering $A < B < C$, is arrowed in the Figure. It is derivable from Euler's equations as they stand:

$$A\dot{\omega}_1 = (B - C)\omega_2\omega_3, \quad B\dot{\omega}_2 = (C - A)\omega_3\omega_1, \quad C\dot{\omega}_3 = (A - B)\omega_1\omega_2. \tag{6}$$

For example, at a point $\omega_2 = 0$ in the A, C plane, $\dot{\omega}_2$ is positive in the first quadrant, negative in the second, etc.

Proceeding, now, to the formal integration of (6) we notice certain other first integrals, dependent on the previous ones but more convenient. By eliminating each component of spin in turn, either between (1) and (2) or between equations (6) taken pairwise,

$$\left.\begin{aligned} B(B - A)\omega_2^2 + C(C - A)\omega_3^2 &= h^2 - 2AT, \\ C(C - B)\omega_3^2 + A(A - B)\omega_1^2 &= h^2 - 2BT, \\ A(A - C)\omega_1^2 + B(B - C)\omega_2^2 &= h^2 - 2CT. \end{aligned}\right\} \tag{7}$$

These could be regarded as the projections of the polhode on the principal planes, one being a full ellipse and the others twice-taken segments of an ellipse and hyperbola respectively (for $A < B < C$, implying $A \leqslant h^2/2T \leqslant C$). As in §3.6 (i) we satisfy one such integral, say the last when the polhode encircles the C axis, by the parametric substitution

$$\left\{\frac{A(A - C)}{h^2 - 2CT}\right\}^{\frac{1}{2}} \omega_1 = \cos\chi, \quad \left\{\frac{B(B - C)}{h^2 - 2CT}\right\}^{\frac{1}{2}} \omega_2 = \sin\chi, \tag{8}$$

where the radicals are real and are arbitrarily taken to be positive. From either remaining integral,

$$\left\{\frac{C(C - A)}{h^2 - 2AT}\right\}^{\frac{1}{2}} \omega_3 = \sqrt{(1 - k^2 \sin^2\chi)}, \quad k^2 = \left(\frac{B - A}{B - C}\right)\left(\frac{h^2 - 2CT}{h^2 - 2AT}\right), \tag{9}$$

where the left-hand radical is again taken positive. The sign of the right-hand radical is not then disposable, being determined by the initial spin (a distinction that the different notation is intended to emphasize). The constant k ranges from 0 to ∞ as $h^2/2T$ ranges from C to A, and is 1 when $h^2/2T = B$; it is also equal to one of these special values when there is axial kinetic symmetry (§3.8). To find the variation of the parameter with time, we can substitute (8) in the first or second of (6), obtaining

$$\left\{\frac{(A-C)(B-C)}{AB}\right\}^{\frac{1}{2}} \omega_3 = \dot{\chi} \tag{10}$$

and then

$$\left\{\frac{(C-B)(h^2-2AT)}{ABC}\right\}^{\frac{1}{2}} t = \int^{\chi} \frac{d\chi}{\sqrt{(1-k^2 \sin^2 \chi)}} \tag{11}$$

by comparing (9) and (10). The integration (Jacobi 1849) involves elliptic functions except when $k = 0$, 1 or ∞. The time of description of a body cone round the C axis is given by (11) when the range of integration is 2π. In this period the instantaneous axis sweeps out a sector of the space cone corresponding to two oscillations between consecutive positions of greatest deviation from the invariable line (cf. the segment of the herpolhode indicated in Fig. 25).

To determine the body's own position in the adopted frame Eulerian coordinates are used to specify the orientation of the principal axes, with C defined by polar angles θ and ϕ with respect to the invariable line (§3.6). Then, by resolving the vector moment of momentum,

$$A\omega_1 = -h \sin\theta \cos\psi, \quad B\omega_2 = h \sin\theta \sin\psi, \quad C\omega_3 = h \cos\theta.$$

Accordingly, once the spin components have been computed as functions of time, the angles θ and ψ are immediately obtainable from

$$\cos\theta = \frac{C\omega_3}{h}, \quad \tan\psi = \frac{-B\omega_2}{A\omega_1} = -\left\{\frac{B(A-C)}{A(B-C)}\right\}^{\frac{1}{2}} \tan\chi. \tag{12}$$

Finally, from the kinematic relations (4) in §3.6 for ω_1 and ω_2,

$$\left(\dot{\phi} - \frac{h}{A}\right) \sin\theta = \dot{\theta} \tan\psi, \quad \left(\dot{\phi} - \frac{h}{B}\right) \sin\theta = -\dot{\theta} \cot\psi,$$

or

$$\dot{\phi} = \frac{h}{A} \cos^2\psi + \frac{h}{B} \sin^2\psi, \tag{13}$$

which is integrated to give ϕ. It is apparent that values of θ and ψ are repeated after an interval equal to the fundamental period of description of the body cone, but that ϕ meanwhile increases by an amount usually incommensurable with 2π. Consequently, the body never occupies a given position more than once, though it returns arbitrarily near in the course of time.

Exceptionally, when $A < h^2/2T = B < C$, the motion is aperiodic since $\dot{\chi}/\cos\chi$ is constant and χ tends either to $\frac{1}{2}\pi$ or $-\frac{1}{2}\pi$ as $t \to \infty$, depending on the initial spin. Correspondingly, both ω_1 and $\omega_3 \to 0$ along a branch of the 'separating polhode' in one of the planes (5). The instantaneous axis approaches the B axis asymptotically, and both therefore tend to align with the invariable direction. It is shown in (iii) below that the planes through the invariable line that contain respectively the instantaneous axis and the B axis rotate at the same constant rate h/B. These planes are in fact orthogonal, since (5) is the condition that their normal directions $\mathbf{h} \wedge \boldsymbol{\omega}$ and $(C\omega_3, 0, -A\omega_1)$ are perpendicular.

(i) *Stability of rotation about a principal direction*

Suppose that a transitory disturbance causes the instantaneous axis to depart from either the direction of least or greatest moment of inertia. In the ensuing free rotation the polhode can be brought arbitrarily near the principal direction in question (say C) by making the disturbance vanishingly small. This is obvious merely from the last integral in (7), here the equation of an ellipse, which implies that ω_1 and ω_2 necessarily retain their joint initial order of smallness. It is clear also from Poinsot's representation that the herpolhode is correspondingly near the invariable line, while both this and the fixed plane are close to their original positions when T and $\mathbf{h}$ are altered infinitesimally. In this sense, then, the original rotation can be described as stable in respect of the axis and magnitude of spin.

To examine the disturbed motion in more detail we linearize the leading pair of Euler equations (6):

$$A\dot{\omega}_1 = (B - C)\nu\omega_2, \quad B\dot{\omega}_2 = (C - A)\nu\omega_1.$$

These are correct to first order in the supposedly infinitesimal ω_1 and ω_2 since, from any of the integrals (1), (2) and (7), ω_3 varies by only a second-order quantity from its disturbed value, ν say. The solution of this simple system is

$$\omega_1/\nu = \varepsilon_1 \cos pt + \alpha\varepsilon_2 \sin pt,$$

$$\omega_2/\nu = \varepsilon_2 \cos pt + \alpha\varepsilon_1 \sin pt,$$

where $$\alpha = \left\{\frac{A(C - A)}{B(C - B)}\right\}^{\frac{1}{2}}, \quad p = \left\{\frac{(C - A)(C - B)}{AB}\right\}^{\frac{1}{2}} \nu,$$

and $\omega_1 = \nu\varepsilon_1$, $\omega_2 = \nu\varepsilon_2$ when $t = 0$, the same sign being taken for each radical. The angular frequency p is real when C is either greater or less than A and B, as expected from the preceding theorem.

When the undisturbed rotation is about the axis B of intermediate moment, it is clear that the second integral in (7) does not constrain ω_1 and ω_3 to remain small after the disturbance, since the coefficients have opposite signs. However, this by itself is not necessarily an indication of instability, and to settle the question we must appeal to the general solution. When the disturbed rotation is not initially about an axis in either of the separating planes, the ensuing polhode is described monotonically around the A or C direction and remains finite however small the disturbance might be (Fig. 25). On the other hand, when the instantaneous axis does initially lie in a separating plane, it either returns immediately towards its undisturbed direction in the body or turns through nearly 180° to the exactly opposed direction. The conclusion is, therefore, that the permanent rotation is unstable under *arbitrary* disturbance.

(ii) *Rotation of a disc-like body*

Consider a body that can be treated as a plane distribution of matter, possibly non-uniform, for which $C = A + B$ (§3.4). One of the integrals (7) becomes symmetrical, namely

$$AB(\omega_1^2 + \omega_2^2) = 2CT - h^2,$$

so that the parametrizations (8) and (10) simplify to

$$\frac{\omega_1}{\cos \chi} = \frac{\omega_2}{\sin \chi} = \left\{\frac{(2CT - h^2)}{AB}\right\}^{\frac{1}{2}}, \qquad \omega_3 = \dot{\chi}. \tag{14}$$

To preserve the symmetry we can write the other independent integral as

$$\omega_3^2 = \frac{(h^2 - 2AT)}{BC} \cos^2 \chi + \frac{(h^2 - 2BT)}{AC} \sin^2 \chi \tag{15}$$

in replacement of (9), obtaining it directly from (1) say. Since the component spin in the plane of the disc has constant magnitude, ω_3 directly determines the angular deviation of the instantaneous axis from this plane.

When, further, $A < B = h^2/2T$ the spin components are simply

$$\omega_1 = \frac{h}{B}\cos\chi, \quad \omega_2 = \frac{h}{B}\sin\chi, \quad \omega_3 = \sqrt{\left(\frac{B-A}{B+A}\right)}\frac{h}{B}\cos\chi,$$

where the radical has the sign of ω_1/ω_3 given initially. Also

$$\tan(\tfrac{1}{2}\chi) = \tanh\left\{\tfrac{1}{2}\sqrt{\left(\frac{B-A}{B+A}\right)}\frac{h}{B}t\right\}$$

where the time origin is arbitrarily taken at the instant when $\chi = 0$ (supposing ω_1 positive). As $t \to \infty$, $\chi \to \frac{1}{2}\pi$ or $-\frac{1}{2}\pi$ according as ω_1/ω_3 is positive or negative initially; correspondingly, ω_1 and $\omega_3 \to 0$ while $\omega_2 \to h/B$ or $-h/B$ (cf. Fig. 25). The disc eventually spins about the B diameter, with this aligned in the invariable direction.

(iii) *Orbital theory of the herpolhode* (*Poinsot*)

The terminus of the spin vector is specified in the fixed plane by polar coordinates (ρ, η) with respect to the point of intersection by the invariable line. Then

$$\rho^2 = \omega^2 - (2T/h)^2 \quad \text{where} \quad \omega^2 = \omega_1^2 + \omega_2^2 + \omega_3^2. \tag{16}$$

By solving this last relation together with (1) and (2) for the spin components,

$$\left.\begin{aligned} (A-B)(A-C)\omega_1^2 &= BC(\rho^2 - \alpha^2), \\ (B-C)(B-A)\omega_2^2 &= CA(\rho^2 - \beta^2), \\ (C-A)(C-B)\omega_3^2 &= AB(\rho^2 - \gamma^2), \end{aligned}\right\} \tag{17}$$

where we introduce the constants

$$\left.\begin{aligned} \alpha^2 &= -\left(\frac{2T}{h} - \frac{h}{B}\right)\left(\frac{2T}{h} - \frac{h}{C}\right), \\ \beta^2 &= -\left(\frac{2T}{h} - \frac{h}{C}\right)\left(\frac{2T}{h} - \frac{h}{A}\right), \\ \gamma^2 &= -\left(\frac{2T}{h} - \frac{h}{A}\right)\left(\frac{2T}{h} - \frac{h}{B}\right). \end{aligned}\right\} \tag{18}$$

With the ordering $A \leqslant B \leqslant C$, β^2 is the greatest and always $\geqslant 0$, but $\alpha^2 \geqslant 0$ only when $B \leqslant h^2/2T \leqslant C$ and $\gamma^2 \geqslant 0$ only when

$A \leqslant h^2/2T \leqslant B$. Consequently just two of α, β and γ are real, one always being β. By multiplying equations (6) in turn by ω_1/A, ω_2/B, ω_3/C, and adding,

$$ABC\omega\dot{\omega} = (C - B)(B - A)(A - C)\omega_1\omega_2\omega_3.$$

By squaring this and substituting from (17),

$$\rho^2\dot{\rho}^2 = (\alpha^2 - \rho^2)(\beta^2 - \rho^2)(\gamma^2 - \rho^2) \tag{19}$$

after using (16). This defines the time variation of the herpolhode radius. The right side is positive only when the radius lies in the annulus bounded above by the positive root β and below by the positive root α or γ. At the outer boundary the component ω_2 vanishes and at the inner boundary either ω_1 or ω_3; the other components thereupon follow from (7).

We prove next that the angular motion is defined by

$$\rho^2\left(\dot{\eta} - \frac{2T}{h}\right) = \left(\frac{2T}{h} - \frac{h}{A}\right)\left(\frac{2T}{h} - \frac{h}{B}\right)\left(\frac{2T}{h} - \frac{h}{C}\right) = i\alpha\beta\gamma. \tag{20}$$

Evidently the herpolhode is a central orbit with respect to a frame turning about the invariable line at rate $2T/h$. From kinematics, considering the rotation of the plane through the invariable line and instantaneous axis,

$$h\rho^2\dot{\eta} = (\mathbf{h} \wedge \boldsymbol{\omega})\dot{\boldsymbol{\omega}} = (\mathbf{h} \wedge \boldsymbol{\omega})J(\mathbf{h} \wedge \boldsymbol{\omega})$$

by the vector form of Euler's equations where J is the inverse of the inertia tensor I (§3.5). But, identically,†

$$|I|(\mathbf{h} \wedge \boldsymbol{\omega})J(\mathbf{h} \wedge \boldsymbol{\omega}) \equiv (\mathbf{h}\, I\mathbf{h})(\boldsymbol{\omega} I \boldsymbol{\omega}) - (\mathbf{h} I \boldsymbol{\omega})^2$$

where $|I| = ABC$ is the determinant of matrix I. Whence

$$ABC\rho^2\dot{\eta} = \mathbf{h}\left(\frac{2T}{h} I - hU\right)\mathbf{h} \tag{21}$$

where U is the unit tensor. This expression is not, of course, constant since I has varying components on fixed axes while $\mathbf{h}$ has varying

† This useful identity is not well known. It is clearly akin, however, to $(\mathbf{a} \wedge \mathbf{b})^2 \equiv \mathbf{a}^2\mathbf{b}^2 - (\mathbf{ab})^2$, from which it can be derived when written *in extenso* in terms of components on the principal axes. The related vector identity

$$|I|J(\mathbf{h} \wedge \boldsymbol{\omega}) \equiv I\mathbf{h} \wedge I\boldsymbol{\omega}$$

is worth noting also.

components on embedded axes. It is evaluated in terms of ω by forming the determinantal eliminant of the spin components between

$$\boldsymbol{\omega} I^3 \boldsymbol{\omega} = \mathbf{h} I \mathbf{h}, \quad \boldsymbol{\omega} I^2 \boldsymbol{\omega} = h^2, \quad \boldsymbol{\omega} I \boldsymbol{\omega} = 2T, \quad \boldsymbol{\omega} U \boldsymbol{\omega} = \boldsymbol{\omega}^2,$$

or more elegantly by drawing on the Cayley–Hamilton theorem that I satisfies its own characteristic equation, whatever the coordinates:

$$I^3 - (A + B + C)I^2 + (AB + BC + CA)I - ABCU = 0.$$

Thus,

$$\mathbf{h} I \mathbf{h} - (A + B + C)h^2 + (AB + BC + CA)2T - ABC\omega^2 = 0. \tag{22}$$

By returning with this to (21), having regard to (16), the required result (20) follows directly.

When $h^2/2T = B$ the herpolhode is not bounded internally since $\alpha = \gamma = 0$. Equations (19) and (20) reduce to

$$\rho \cosh (\beta t) = \beta, \qquad \eta = \frac{h}{B} t,$$

where

$$\beta = \frac{h}{B}\left\{\frac{(C - B)(B - A)}{AC}\right\}^{\frac{1}{2}}.$$

The time origin and the η base line have here been chosen to coincide with the tangency on the outer boundary. The locus consists of two identical spirals reflected in this line and described during positive or negative times respectively.

3.8 Steady States: Precession

The motion of a rigid body relative to one of its particles P is said to be a steady or regular precession when some embedded line through P generates a circular cone uniformly and the component spin of the body about this line is constant. Naturally, the motion relative to any other particle is then also of this kind. The kinematics of precession are now examined and afterwards the dynamics.

Let $\mathbf{k}$ be a unit vector in the fixed direction of the axis of precession, and $\mathbf{l}$ a unit vector in the varying direction of the embedded line. Let ν be the component spin of the body resolved on the line $\mathbf{l}$, and Ω the angular rate at which this line turns right-handedly about the axis $\mathbf{k}$. Then, by (4) in §3.1 with $\dot{\mathbf{l}} = \mathbf{\Omega} \wedge \mathbf{l}$, $\mathbf{\Omega} \equiv \Omega\mathbf{k}$, the spin of the body is

$$\boldsymbol{\omega} = \mathbf{l} \wedge (\mathbf{\Omega} \wedge \mathbf{l}) + \nu\mathbf{l}. \tag{1}$$

This decomposes the spin into orthogonal components ν and $\Omega \sin \alpha$ in the axial plane, where α is the inclination of the embedded line. Expanding the product and putting $\mathbf{kl} = \cos \alpha$,

$$\boldsymbol{\omega} = \Omega\mathbf{k} + (\nu - \Omega \cos \alpha)\mathbf{l}. \tag{2}$$

This is an oblique decomposition, otherwise evident from the standpoint of an auxiliary frame with spin $\mathbf{\Omega}$. Since α is constant the

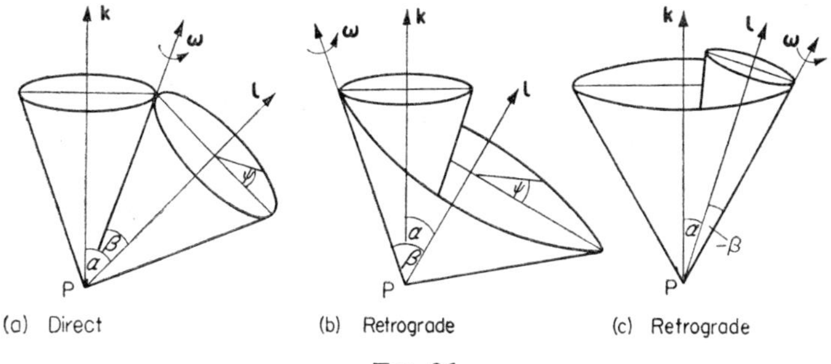

FIG. 26

embedded line is at rest in that frame and is therefore the instantaneous axis of the apparent motion $\boldsymbol{\omega} - \mathbf{\Omega}$. This is then of form $\lambda\mathbf{l}$ where $\lambda = (\boldsymbol{\omega} - \mathbf{\Omega})\mathbf{l}$ by resolution, as in (2).

Since $\boldsymbol{\omega}$ is coplanar with $\mathbf{k}$ and $\mathbf{l}$, and has a fixed inclination to both, the space and body cones for the motion about P are circular. Their semi-vertex angles are denoted by $\alpha - \beta$ and β respectively (Fig. 26a). Moreover, since the spin magnitude ω is constant, the polhode and herpolhode curves are also circular. Specifically, by resolving (1) along and transversely to the embedded line,

$$\omega \cos \beta = \nu, \qquad \omega \sin \beta = \Omega \sin \alpha, \tag{3}$$

from which the spin vector is determinable when the precessional constants Ω, ν, and α are given. The cones are either mutually convex as in (*a*) or not, as in (*b*) and (*c*); but the same formulae apply in all cases provided that β is reckoned as a negative angle in (*c*) when Ω and ω have opposite signs. Conversely, if the time variation of $\boldsymbol{\omega}$ is prescribed such that the herpolhode is a circle uniformly described about $\mathbf{k}$, the body must precess steadily about P. Further, the one embedded line through P that preserves its inclination to the axis of precession has the direction of $\boldsymbol{\omega} - \boldsymbol{\Omega}$.

The body orientation is completely specified by the direction of the embedded line and the Eulerian angle ψ measured right-handedly round it from the plane of $\mathbf{k}$ and $\mathbf{l}$ (cf. Fig. 23). Then

$$\left.\begin{aligned} \boldsymbol{\omega} &= \Omega\mathbf{k} + \dot{\psi}\mathbf{l} \\ \text{where} \qquad \dot{\psi} &= \nu - \Omega\cos\alpha = \Omega\sin(\alpha - \beta)/\sin\beta \end{aligned}\right\} \tag{4}$$

from (3) or by resolving transversely to the instantaneous axis. It is seen that $\dot{\psi}$ and Ω have the same sign in (a) where $\alpha > \beta$ but the opposite sign in (b) where $\beta > \alpha$ or in (c) where β is negative. The precession is correspondingly said to be **direct** or **retrograde**.

Suppose that the line $\mathbf{l}$ is an axis of kinetic symmetry and that C is the moment of inertia. Let P be any point on this axis and A the moment of inertia about every transverse direction there. Then the moment of *relative* momentum about P is

$$A\mathbf{l} \wedge (\boldsymbol{\Omega} \wedge \mathbf{l}) + C\nu\mathbf{l} = A\mathbf{l} \wedge (\boldsymbol{\omega} \wedge \mathbf{l}) + C\nu\mathbf{l},$$

by reference to the decomposition (1). Equivalent expressions are

$$A\boldsymbol{\Omega} + (C\nu - A\Omega\cos\alpha)\mathbf{l} \tag{5}$$

and

$$A\boldsymbol{\omega} + (C - A)\nu\mathbf{l},$$

by expanding the products. When, more particularly, P is either the mass-centre or a fixed pivot, these are also expressions for the moment of *absolute* momentum, $\mathbf{h}$. In that event their rate of change determines the resultant 'gyroscopic' couple $\mathbf{M}$ about P needed to maintain the precession, once started. Since $\mathbf{h}$ is here a fixed vector in the rotating auxiliary frame, $\mathbf{M} = d\mathbf{h}/dt = \boldsymbol{\Omega} \wedge \mathbf{h}$ simply or

$$\mathbf{M} = (C\nu - A\Omega\cos\alpha)\boldsymbol{\Omega} \wedge \mathbf{l} \tag{6}$$

from (5). The axis of the couple is thus perpendicular to the axial plane and turns with it. When the body has all-round kinetic symmetry at P, so that $C = A$, an alternative to (6) is

$$\mathbf{M} = A\boldsymbol{\Omega} \wedge \boldsymbol{\omega} \tag{7}$$

since $\mathbf{h} = A\boldsymbol{\omega}$. This variant is useful when the particular constraints present are such that $\boldsymbol{\omega}$ can be recognized more easily than $\mathbf{l}$.

The *free* rotation of an axially symmetric body about its mass-centre is always a precession. This is already clear from Poinsot's representation (§3.7) since the herpolhode traced on the constraining plane by the rolling spheroid can only be a circle, necessarily described uniformly. But the constants of precession can not be assigned at will by choice of the initial conditions; in fact by (6) they are related by

$$C\nu = A\Omega \cos \alpha. \tag{8}$$

For instance, if the orientation of the body together with the spin vector are given initially, ω, ν and β are known and then (3) with (8) are to be regarded as equations for Ω and α determining the axis and rate of precession:

$$\Omega \cos \alpha = \frac{C}{A} \omega \cos \beta, \qquad \Omega \sin \alpha = \omega \sin \beta. \tag{9}$$

The vector moment of momentum, now directed along the axis, depends only on the rate of precession since (5) reduces to

$$\mathbf{h} = A\boldsymbol{\Omega}. \tag{10}$$

Again, by dividing the equations (9), $C \tan \alpha = A \tan \beta$ which shows that case (c) in Fig. 26 does not arise here (this is also evident by considering that the radial and normal directions $\boldsymbol{\omega}$ and $\mathbf{h}$ to the inertia spheroid can not fall on opposite sides of its axis of symmetry). The precession is direct when $C < A$, retrograde when $A < C \leqslant 2A$ (disc). Finally, from (4) and (8), the rate of description of the body cone relatively to the material is

$$-\dot{\psi} = \left(\frac{C}{A} - 1\right) \nu \tag{11}$$

right-handedly about the C axis. All these results follow alternatively from the general theory in §3.7, which shows to begin with that the polhodes are circles about the axis of symmetry:

$$\omega_1^2 + \omega_2^2 = \Omega^2 \sin^2 \alpha, \qquad \omega_3 = \nu,$$

from the integrals (7) there, with the right-hand constants written in the present notation. It is best to derive the succeeding equations in §3.7 *ab initio*, avoiding separate treatment of the cases $A = B \gtrless C$.

Thus, by introducing the parametrization $\omega_1/\cos\chi = \omega_2/\sin\chi = \Omega\sin\alpha$ in Euler's equations, we obtain $A\dot{\chi} = (C-A)\nu$ directly and can then identify χ with $-\psi$ by comparison with (11).

Consider, next, a body having an axis of kinetic symmetry and pivoted at a point O at a distance d along this axis from the mass-centre. Steady precession about the vertical under its own weight is possible since the available couple about O then has the direction and constant magnitude required by (6) provided $M = mgd\sin\alpha$, or

$$A\Omega^2\cos\alpha - C\nu\Omega + mgd = 0 \tag{12}$$

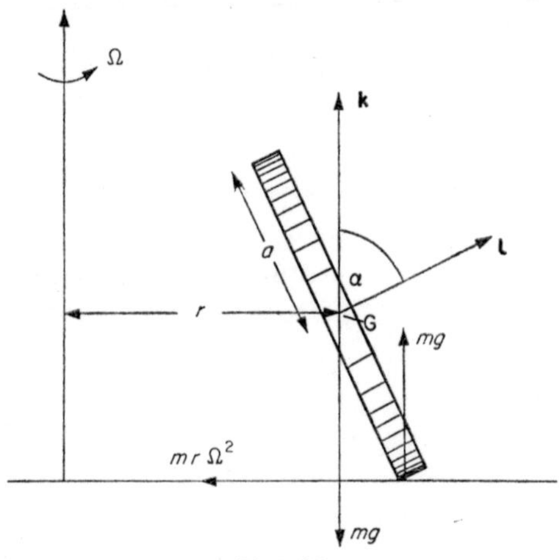

FIG. 27

where A is the transverse principal moment at O. If, for example, the inclination α is assigned, the motion has to be started with a sufficiently fast component spin ν when $\alpha < \frac{1}{2}\pi$, namely such that $C^2\nu^2 \geqslant 4Amgd\cos\alpha$, and with a rate of precession satisfying (12). There are two possible rates of right-handed precession about the upward vertical with $\alpha < \frac{1}{2}\pi$; when ν is very large these are $mgd/C\nu$ and $C\nu/A\cos\alpha$, approximately. When $\alpha > \frac{1}{2}\pi$ (gyroscopic pendulum) only the slower precession is right-handed. An idealized rod ($C = 0$) necessarily rotates as a pendulum and the two precessions differ only in sense.

(i) *Rolling disc*

A disc of radius a rolls on a horizontal plane to which it maintains an angle α (Fig. 27). Suppose that its centre G travels in a circle of radius r with speed $r\Omega$, where Ω is also the rate of precession of the disc about G. Since the particle at the point of contact has speed av relative to G and in the opposite sense, it is instantaneously at rest only if $v = -r\Omega/a$. The forces available at the point of contact to supply the gyroscopic couple about G are the upward reaction mg and the inward friction $mr\Omega^2$, these being necessary for the circular motion of G. Then, from (6) applied at G with $2A = C = mk^2$,

$$\left\{\left(1+\frac{k^2}{a^2}\right)\frac{r}{a}+\frac{k^2\cos\alpha}{2a^2}\right\}\Omega^2 = \frac{g}{a}\cot\alpha.$$

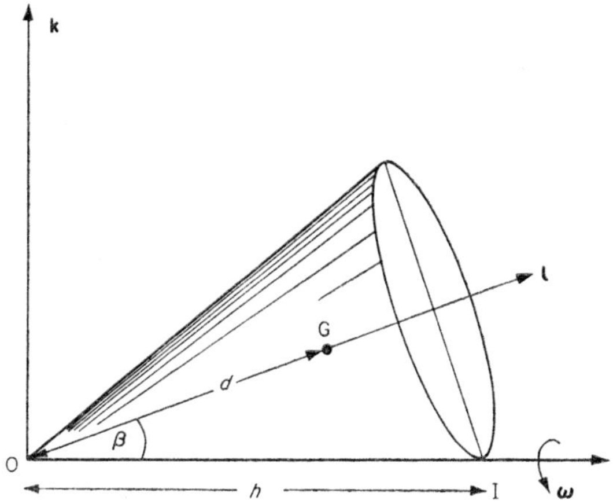

FIG. 28

This may be read as an equation determining any one of Ω, r, and α in terms of the other two.

(ii) *Cone rolling on a plane*

A homogeneous right circular cone with vertex angle 2β and slant height h rolls on a fixed horizontal surface (Fig. 28); it is required to find the greatest possible rate of precession. This is an example of the second type of retrograde motion since the instantaneous axis is here the generator OI in contact. Therefore, by reversing the sign of β in (3) and then putting $\alpha = \frac{1}{2}\pi - \beta$,

$$v \tan\beta = -\Omega\cos\beta$$

is the expression of the rolling constraint. When multiplied by the distance $OG = d$, the two sides are the circumferential speed of the mass-centre G when regarded respectively as a particle of the rolling cone and as a point in the precessing plane. The gyroscopic couple at the stationary point O is thus

$$(C \cos^2 \beta + A \sin^2 \beta)\Omega^2 \cot \beta$$

anticlockwise, by substituting for ν and α in (5). The bracketed quantity is evidently the moment of inertia about a generator, mk^2 say. Its appearance can be understood by reflecting that the convected part of the moment of momentum, namely the component perpendicular to the axis of precession, is simply $mk^2\omega$ when the

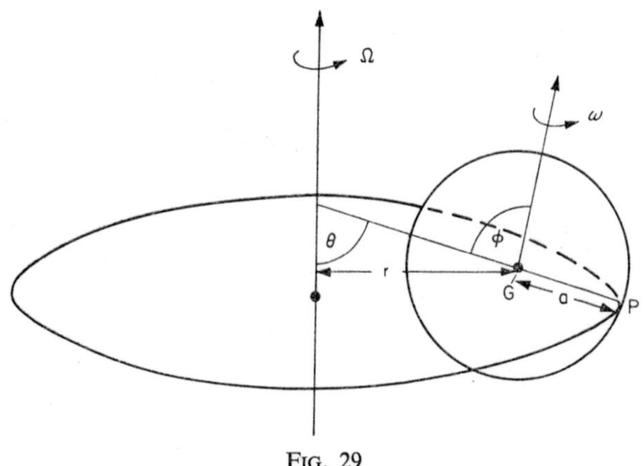

FIG. 29

spin vector is itself perpendicular, as here. The couple is accordingly $-mk^2\Omega\omega$, with $\omega = -\Omega \cot \beta$.

This can only be supplied by the weight mg acting through G and the normal reactions distributed along OI, the friction making no contribution. Since the reactions are necessarily pressures with linear resultant mg their moment about O does not exceed mgh and attains this value when the cone is on the verge of pivoting about I. Consequently, full contact is retained if and only if

$$k^2\Omega^2 < g(h - d \cos \beta) \tan \beta,$$

yielding the upper bound on Ω.

(iii) *Sphere rolling on a wire circle*

A centro-symmetric sphere with radius a and diametral moments of inertia mk^2 rolls round a horizontal wire circle (Fig. 29). The diameter from the point of contact P is inclined at θ to the vertical and the motion is started so that the spin vector is inclined at ϕ to this diameter. If Ω is the rate of precession and ω the body spin, with the senses indicated, the rolling constraint is

$$r\Omega + a\omega \sin \phi = 0$$

where r is the radius of the orbit of the mass-centre G. In this situation formula (7) is most convenient since the embedded line with constant inclination is not known *a priori* (it is not GP because different particles successively occupy the position P). The rotational equation of balance about G is then

$$k^2\Omega\omega \sin (\phi - \theta) = a(r\Omega^2 \cos \theta - g \sin \theta)$$

since the contact reaction has components mg upward and $mr\Omega^2$ horizontally inward. Eliminating ω:

$$\left\{\left(1 + \frac{k^2}{a^2}\right) \cot \theta - \frac{k^2}{a^2} \cot \phi\right\} r\Omega^2 = g,$$

provided the configuration is such that the expression is positive.

(iv) *Degenerate precession*

When the instantaneous axis coincides with the axis of precession the motion reduces to a pure rotation $\boldsymbol{\omega} = \omega\mathbf{k}$ about this fixed direction, to which every embedded line then maintains a constant inclination. For a body with axial symmetry $\beta = \alpha$, $\Omega = \omega$, $\nu = \omega \cos \alpha$, $\dot{\psi} = 0$ in the previous notation and $M = (C - A)\omega^2 \sin \alpha \cos \alpha$ about the normal to the plane of $\mathbf{k}$ and the C direction. For a general body the couple $\mathbf{M} = \omega\mathbf{k} \wedge \mathbf{h}$ is also perpendicular to $\mathbf{k}$, with components $(C - B)k_2k_3\omega^2$, etc. on the principal axes at the mass-centre G or at a particle O permanently at rest. The preceding result is recovered when $A = B$, $\mathbf{k} \equiv (-\sin \alpha, 0, \cos \alpha)$, $\mathbf{M} \equiv (0, M, 0)$.

When the body is mounted on a fixed *shaft* and is not under externally applied forces, the considered uniform spin is possible if G is situated on the shaft and this is frictionless. In that event the induced normal reactions can provide the needed couple whatever the orientation of the shaft within the body. When, however, the body is free to turn about a *pivot* at G the axis of permanent rotation can only be a principal direction. But when the pivot is not at G a different and non-trivial configuration becomes possible (Staude 1894).

In the first place any permanent axis of rotation must be vertical so that G may describe a horizontal circle, otherwise the weight would have a varying lever arm about the pivot O. In the second place the radius from O to G, $d\mathbf{l}$ say, must be coplanar with $\mathbf{k}$ and $\mathbf{h}$ so that the weight may have a moment whose axis is normal to that plane. That is, $\mathbf{l}(\mathbf{k} \wedge \mathbf{h}) = 0$ or

$$l_1(B - C)k_2k_3 + l_2(C - A)k_3k_1 + l_3(A - B)k_1k_2 = 0$$

in terms of components on the principal axes at O. This defines in the body a quadric cone of permissible directions $\mathbf{k}$, one of which must be set vertically. This having been selected, the vector equation of balance $\omega\mathbf{k} \wedge \mathbf{h} = mgd\mathbf{k} \wedge \mathbf{l}$ is satisfied with a rate of rotation given by any one of

$$\frac{\omega^2}{mgd} = \frac{l_2k_3 - l_3k_2}{(B - C)k_2k_3} = \frac{l_3k_1 - l_1k_3}{(C - A)k_3k_1} = \frac{l_1k_2 - l_2k_1}{(A - B)k_1k_2},$$

which are obtained by resolving on the principal directions. The sense of $\mathbf{k}$ is not arbitrary, but such that the quotients are positive.

Any generator of the Staude cone is a possible vertical axis of permanent rotation provided these equations define a proper spin. They fail to do so for the generator OG since $\omega = 0$ when $\mathbf{k} = \mathbf{l}$ (unless, trivially, this direction is principal and so coincident with $\mathbf{h}$), and also for each principal direction at O (unless, again, it passes through G) since $\omega \to \infty$ when $\mathbf{k} \to (1, 0, 0)$ for instance. However, together with

$$\mathbf{k} = \frac{mgd}{\omega^2}\left(\frac{l_1}{A}, \frac{l_2}{B}, \frac{l_3}{C}\right),$$

these five special generators can be regarded as completely defining the Staude cone. This becomes a pair of planes when OG lies in a principal plane, this itself being one of the pair. The cone is non-existent when OG is an axis of kinetic symmetry, since all coefficients in its equation vanish when $A = B$ and $l_1 = l_2 = 0$, $\mathbf{h}$ being automatically coplanar with any $\mathbf{k}$ and $\mathbf{l}$. The spin is therewith defined by $(C - A)k_3\omega^2 = mgd$ which agrees with (12) when $\Omega = \omega$, $\cos\alpha = k_3$, $\nu = k_3\omega$ [cf. §3.6 (ii)].

On the axis of rotation a point P can be found about which the acting forces have no resultant moment. For, with a the depth of P below the pivot and r the orbital radius of G, the moment ($a \times mr\omega^2$) of the horizontal component of the reaction at O annuls the moment ($r \times mg$) of the weight if $\omega^2 = ga$. The vector moment of momentum about P, which is at rest, is therefore constant in time and so directed along the axis of rotation (otherwise there would be a convection effect). Consequently, this axis must be a principal direction at P. In other words the Staude cone has the characteristic that every generator is principal at some point. This can alternatively be shown via the formal properties of the inertia tensor, noting that the plane of **k** and **h** contains G (cf. Exercise 11).

(v) *Precession of the earth*

Unless a planet is centrobaric its gravitational action on an external mass-point P' is generally not directed towards its own mass-centre P. In this event the equal and opposite action of the mass-point has a moment **M** about P given in gravitational units by $\mathbf{r} \wedge (-\mu'\mathbf{f})$ where $\mathbf{r}$ is the radius vector from P to P' and μ' is the gravitational mass of P'. With sufficient accuracy at great distances **f** can be taken as the explicit part of (6) in §1.16. Then

$$\mathbf{M} = \frac{3\mu'}{2r^3}\,\mathbf{r} \wedge \operatorname{grad}\left(\frac{\mathbf{r}I\mathbf{r}}{r^2}\right) = \frac{3\mu'}{r^5}\,\mathbf{r} \wedge I\mathbf{r}$$

where I is the inertia tensor at P and the bracketed quantity is the moment of inertia about PP' (**M** and I can now be in c.g.s. units). When μ' is an extended body with mass-centre P' the expression clearly remains the leading term in an expansion in inverse powers of the distance.

Suppose the considered planet is the earth, with equatorial and polar moments of inertia A and C. Then the couple is

$$\frac{3\mu'}{r^3}(C - A)\sin\lambda\cos\lambda$$

about an axis normal to the meridian plane through the body, where λ is its astronomical latitude, and in the sense tending to rotate the equatorial plane towards the line of centres. As the sun orbits the earth in the plane of the ecliptic its latitude oscillates between $\pm e$, where e is the obliquity. The average couple due to the sun is easily shown to be half the maximum and about the same axis,

$$\frac{3\mu_S}{2r_S^3}(C - A)\sin e\cos e$$

in magnitude, where μ_S and r_S are its mass and mean distance. A similar formula is written for the average couple due to the moon in terms of its mass μ_M and mean distance r_M. As pointed out by Newton, the observed precession about the ecliptic pole, once every 25,735 years, can only be maintained by this luni-solar couple, the planetary contributions being infinitesimal by comparison. In the present analysis, due to Euler, various small periodic deviations from a steady state are disregarded; these are the so-called 'nutations', both free and forced.† Relative to a sidereal frame the earth makes $366\frac{1}{4}$ rotations a year, very nearly; the rate of precession is therefore less by a factor $25{,}735 \times 366{\cdot}25 = 0{\cdot}9425 \times 10^7$. Consequently, from (3) with $\alpha = e \simeq 23\frac{1}{2}°$ (long-term average), the body cone is utterly insignificant and ω is indistinguishable from ν, while (6) is near enough $-C\Omega\nu \sin e$ with the sense and axis of the luni-solar couple. Ω and ν have, indeed, opposite signs and the precession is retrograde, Fig. 26(c). Whence

$$\frac{C-A}{C} = \frac{\frac{2}{3}\left|\frac{\Omega}{n}\right|\left|\frac{\nu}{n}\right| \sec e}{1 + \frac{\mu_M}{\mu_S}\left(\frac{r_S}{r_M}\right)^3}$$

where $2\pi/n$ is the sidereal period of the earth's orbit given by $\mu_S = n^2 r_S^3$, as in (6) of §1.4.

With data in §1.7,

$$\left|\frac{\nu}{n}\right| = 366{\cdot}25, \quad \left|\frac{\Omega}{n}\right| = \frac{1}{25{,}735},$$

$$\frac{2}{3}\sec e = 0{\cdot}7270, \quad \frac{\mu_M}{\mu_S} = 37{\cdot}03 \times 10^{-9}, \quad \frac{r_S}{r_M} = 0{\cdot}3890 \times 10^3,$$

† For a more exact treatment elaborated by Laplace see E. J. Routh: *Treatise on the Dynamics of a System of Rigid Bodies*, Advanced Part, Chapter XI (1905 edition re-issued by Dover Publications 1955) and A. Gray: *Treatise on Gyrostatics and Rotational Motion*, Chapters X and XI (1918 edition re-issued by Dover Publications 1959).

which leads to

$$\frac{C-A}{C} = \frac{0{\cdot}01035}{1 + 2{\cdot}178} = 0{\cdot}00326.$$

The moon's contribution is thus a little more than twice the sun's. In fact the calculation is quite sensitive to the value assumed for the mean radius of the moon's orbit, and the figure for the 'mechanical ellipticity' is uncertain by about one unit in the last place of decimals. To obtain C and A separately we recall the ratio

$$\frac{C-A}{mr_0^2} = 0{\cdot}00109$$

obtained from the internal and external field theories [§1.18, equation (3) and subsequent data], where m is the inertial mass of the earth and r_0 its mean radius. By dividing the last two relations the radius of gyration about the polar axis is found as $r_0/\sqrt{3}$, very nearly.

3.9 Gyrodynamics

Under this heading an introductory account is given of non-steady rotations of any axially symmetric body, customarily referred to simply as a 'top'. More particularly, when in the form of a flywheel and spindle (the mass being concentrated towards the rim to make C/m and C/A as large as practicable) it is called a 'gyrostat' (Kelvin's terminology, to mark the role of axial spin in tending to keep this direction 'stationary'). When intended primarily as a sensing device, usually mounted so that it can turn freely about the mass-centre in either two or three degrees of freedom, it is called a 'gyroscope' (Foucault's terminology, literally 'to see rotation').† We begin with an analysis of the 'heavy top': this is pivoted not at its mass-centre but at some other axial point, and is otherwise unconstrained. The general theory is largely due to Lagrange (1788) and Poisson (1833).

The orientation in a terrestrial frame of reference is conveniently defined by Eulerian angles θ, ϕ, ψ in respect to the C axis and the upward vertical z at the fixed pivot O (cf. Fig. 23). However, because

† The functioning of a variety of instruments and devices is explained analytically by R. N. Arnold and L. Maunder: *Gyrodynamics and its Engineering Applications* (Academic Press 1961) and A. Gray: *Gyrostatics and Rotational Motion*, Chapters I and VII (1918 edition re-issued by Dover Publications 1959).

of the symmetry, the angle ψ is of no interest and in fact only its derivative enters the vector equation of balance. Accordingly we resolve this on the auxiliary axes (s, n, C) where n is normal to the plane of C and z (Fig. 30), instead of on embedded axes as in §§3.6 and 3.7. Since any direction transverse to C is principal, the moment

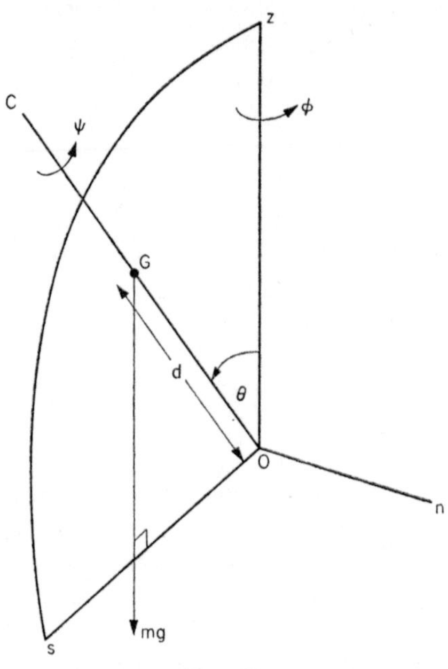

FIG. 30

of momentum $\mathbf{h}$ still has simple components. Thus, if A is the transverse moment of inertia at O and $\boldsymbol{\omega}$, $\boldsymbol{\Omega}$ the spin of the body and the auxiliary axes respectively,

$$\boldsymbol{\Omega} \equiv (-\dot{\phi} \sin \theta, \dot{\theta}, \dot{\phi} \cos \theta), \quad \boldsymbol{\omega} \equiv (-\dot{\phi} \sin \theta, \dot{\theta}, \nu),$$
$$\mathbf{h}(O) \equiv (-A\dot{\phi} \sin \theta, A\dot{\theta}, C\nu), \qquad \nu = \dot{\phi} \cos \theta + \dot{\psi}.$$

The rate of change of $\mathbf{h}$ relative to the auxiliary frame is therefore

$$(-A\ddot{\phi} \sin \theta - A\dot{\theta}\dot{\phi} \cos \theta, \quad A\ddot{\theta}, \quad C\dot{\nu})$$

resolved on the directions (s, n, C). To these components must be added convective terms

$$\boldsymbol{\Omega} \wedge \mathbf{h} \equiv (C\nu - A\dot{\phi} \cos \theta)(\dot{\theta}, \dot{\phi} \sin \theta, 0)$$

to obtain the rate of change relative to the terrestrial frame (§3.1). The corresponding equation of balance expresses its equivalence with the moment of the weight, $mgd \sin \theta$ about On, when the Coriolis couple associated with the earth's rotation and the frictional torque at the pivot are both negligible by comparison. Whence

$$\ddot{\phi} \sin \theta + 2\dot{\theta}\dot{\phi} \cos \theta = \frac{C\nu}{A} \dot{\theta}, \tag{1}$$

$$\left(\dot{\phi}^2 \cos \theta - \frac{C\nu}{A} \dot{\phi} + \frac{mgd}{A}\right) \sin \theta = \ddot{\theta}, \tag{2}$$

together with $\nu =$ constant. It is worth observing that the differential equations in θ and ϕ involve the various parameters in two groups only: $C\nu/A$ and mgd/A. The property of constant axial spin may also be seen vectorially. Since $C\nu = \mathbf{hl}$, where $\mathbf{l}$ is a unit axial vector, $C\dot{\nu} = \dot{\mathbf{h}}\mathbf{l} + \mathbf{h}\dot{\mathbf{l}}$; but $\dot{\mathbf{h}}\mathbf{l} = 0$ since there is no axial couple, and $\mathbf{h}\dot{\mathbf{l}} = \mathbf{h}(\boldsymbol{\omega} \wedge \mathbf{l}) = 0$ since $\mathbf{h}$ is coplanar with $\boldsymbol{\omega}$ and $\mathbf{l}$.

Two more independent first integrals follow readily. Multiplication of (1) by $\sin \theta$ converts it into the perfect differential of

$$\dot{\phi} \sin^2 \theta + \frac{C\nu}{A} \cos \theta = \text{const.}, D \text{ say}. \tag{3}$$

This is recognizably proportional to the component moment of momentum about the fixed vertical at O, and its constancy could have been stated by first principles. It is apparent also that the total energy is conserved:

$$\tfrac{1}{2}(\dot{\theta}^2 + \dot{\phi}^2 \sin^2 \theta) + \frac{mgd}{A} \cos \theta = \text{const.}, E \text{ say}, \tag{4}$$

where the term $C\nu^2/2A$ has been absorbed in E. This is otherwise obtainable by integrating the sum of (1) $\times\, \dot{\phi} \sin \theta$ and (2) $\times\, \dot{\theta}$. The formal analogy between these integrals and those for a sphere rolling on a spherical surface has already been remarked [§3.3 (ii)], the correspondences being

$$\frac{C}{md} = \frac{k^2}{a}, \qquad \frac{A}{md} = \left(1 + \frac{k^2}{a^2}\right)(b - a).$$

Further integration generally introduces elliptic functions. However, the main features of the motion can be established independently. To begin with, the range of inclination is determined by

$$\Psi(\theta) \equiv \frac{1}{2}\left(\frac{D - \frac{C\nu}{A}\cos\theta}{\sin\theta}\right)^2 + \frac{mgd}{A}\cos\theta \leqslant E, \tag{5}$$

since

$$\tfrac{1}{2}\dot{\theta}^2 + \Psi(\theta) = E \tag{6}$$

when the variable rate of precession is eliminated between (3) and (4). As in the theory of a spherical pendulum (§2.5) Ψ is the potential of the body forces in a frame turning round the vertical with plane OCz. For given initial data specifying D and E, the inequality defines a single annulus on an imagined sphere of radius d say. This annulus is a complete cap containing the highest point $\theta = 0$ when and only when $D = C\nu/A$ with $E \geqslant mgd/A$, and the lowest point $\theta = \pi$ when and only when $D = -C\nu/A$ (whatever the initial conditions E is automatically $\geqslant -mgd/A$).

Only if the top is trivially set rotating permanently about the downward vertical does it happen that $D = -C\nu/A$ with $E = -mgd/A$. By contrast $D = C\nu/A$ with $E = mgd/A$ corresponds to another motion besides a 'sleeping top' ($\theta \equiv 0$). Removing the factor $(1 - \cos\theta)$ we find that (5) is satisfied in the cap $\cos(\theta/2) \geqslant C\nu/2\sqrt{(Amgd)}$ when $C\nu/A < 2\sqrt{(mgd/A)}$. If started obliquely the top spirals away from or towards the vertical without ever becoming upright. In fact from (3) we see that the precession is monotonic, with $\dot{\phi} = C\nu/A(1 + \cos\theta)$ one-signed. For small θ the spiral is given parametrically by

$$\phi - \phi_0 \simeq \frac{C\nu}{2A}t, \qquad \frac{\theta}{\theta_0} \simeq \exp\left\{\pm\sqrt{\left(\frac{4Amgd}{C^2\nu^2} - 1\right)}\frac{C\nu}{2A}t\right\},$$

obtained by expanding (6) to second order.

For initial data such that $D/E = C\nu/mgd$ with $D < C\nu/A$ the orbit is cusped at the upper boundary of the annulus and tangential at the lower (the intervening segments being of course identical apart from reflection). At a cusp the axis is momentarily stationary, the stated relations being necessary and sufficient that $\dot{\theta} = \dot{\phi} = 0$ for a certain $\theta \neq 0, \pi$. This happens only at the upper boundary since, by (4), the associated value EA/mgd of $\cos\theta$ is obviously not exceeded elsewhere in the annulus. Further, on dividing (4) by (3), the rate of precession at a boundary is found as

$$\dot{\phi} = \frac{2\left(E - \dfrac{mgd}{A}\cos\theta\right)}{D - \dfrac{Cv}{A}\cos\theta};$$

in particular, this becomes $2mgd/Cv$ at the lower boundary of a cusped orbit, no matter what its level. Conversely, if $\dot{\phi}$ has this particular boundary value, the initial data must necessarily be such that $D/E = Cv/mgd$; the orbit is cusped if the latter ratio is less than unity but not otherwise (in which case $\dot{\phi}$ takes the same value on both boundaries). Thus, when a spinning top is gently released in an inclined position, $\theta = \alpha$ say, this is necessarily the highest point of a cusped orbit. More especially, if the axial spin is large, the resulting range in θ is $O(v^{-2})$ by inspection of (5), and (3) can be approximated by

$$\dot{\phi} \simeq \frac{Cv}{A\sin\alpha}\varepsilon \quad \text{where} \quad \theta = \alpha + \varepsilon(t).$$

Whence, by linearizing (2) similarly and integrating with the initial conditions ε, $\dot{\varepsilon} = 0$ at $t = 0$,

$$\varepsilon = \frac{Amgd\sin\alpha}{C^2v^2}\left\{1 - \cos\left(\frac{Cv}{A}t\right)\right\},$$

$$\phi = \frac{mgd}{Cv}\left\{t - \frac{A}{Cv}\sin\left(\frac{Cv}{A}t\right)\right\}.$$

This represents a harmonic oscillation of high frequency Cv/A superimposed on a slow precession at rate mgd/Cv. This motion is described as 'pseudo-regular' since it is visually indistinguishable from a steady state when the spin is large enough.

When $D = Cv/A$ with $E > mgd/A$ the orbit passes repeatedly through $\theta = 0$ since $\dot{\theta}$ does not vanish there by (4), while $\dot{\phi} = Cv/2A$ there by (3). Similarly, when $D = -Cv/A$ with $E > -mgd/A$ the orbit is a succession of loops through $\theta = \pi$, where $\dot{\phi} = -Cv/2A$. Loops are also obtained in an annulus not containing either pole when $|D| < Cv/A$ together with $E/mgd > D/Cv$. For then (3) and

(4) permit a value of $\theta \neq 0$ or π, namely $\cos^{-1}(DA/C\nu)$, where $\dot{\phi}$ vanishes and changes sign but $\dot{\theta}$ does not.

(i) *Stability of steady precession*

According to the theory of orbital stability in §2.5, precession with inclination α_0 is stable if and only if $\Psi(\theta) - \Psi(\alpha)$ is a positive-definite function of θ near each α in some interval containing α_0, when D is assigned its value for the neighbouring precession with inclination α and the axial spin ν is supposed unchanged. Now, from (5) and (6),

$$\ddot{\theta} + \Psi'(\theta) = 0$$

with $\sin^3\theta\Psi'(\theta) = \left(D - \frac{C\nu}{A}\cos\theta\right)\left(\frac{C\nu}{A} - D\cos\theta\right) - \frac{mgd}{A}\sin^4\theta.$

By direct verification or comparison with (2),

$$A\Psi'(\alpha) \equiv -(A\Omega^2\cos\alpha - C\nu\Omega + mgd)\sin\alpha = 0$$

as in (12) of §3.8, where Ω is the rate of steady precession that would correspond via (3) to $D(\alpha)$:

$$\left.\begin{aligned} \Omega\sin^2\alpha + \frac{C\nu}{A}\cos\alpha = D, \\ \left(D - \frac{C\nu}{A}\cos\alpha\right)\left(\frac{C\nu}{A} - D\cos\alpha\right) = \frac{mgd}{A}\sin^4\alpha. \end{aligned}\right\} \tag{7}$$

Further,

$$\sin^2\alpha\Psi''(\alpha) = \left(D - \frac{C\nu}{A}\cos\alpha\right)^2 + \left(\frac{C^2\nu^2}{A^2} - \frac{4mgd}{A}\cos\alpha\right)\sin^2\alpha,$$

which is always positive since a steady state only exists when

$$(C\nu - 2A\Omega\cos\alpha)^2 = C^2\nu^2 - 4Amgd\cos\alpha \geqslant 0. \tag{8}$$

Precession is therefore certainly stable and the angular frequency p of small oscillations in θ and $\phi - \Omega t$ is such that

$$p^2 = \Psi''(\alpha) = \Omega^2\sin^2\alpha + \left(\frac{C^2\nu^2}{A^2} - \frac{4mgd}{A}\cos\alpha\right) \tag{9}$$

where, after the disturbance, α is the mean inclination and Ω the corresponding rate of precession. This expression can be given several forms by use of (7) and (8).

A sleeping top needs separate examination. The equation of balance is most conveniently linearized in regard to motion near the pole $\theta = 0$ by starting with the variant

$$\frac{d}{dt}(A\mathbf{l} \wedge \dot{\mathbf{l}} + Cv\mathbf{l}) = md\mathbf{l} \wedge \mathbf{g}$$

where $\mathbf{l}$ is a unit axial vector. Resolving on the horizontal and forming the complex quantity ζ from the corresponding projections of $\mathbf{l}$ on some rectangular axes, we obtain

$$A\ddot{\zeta} - iCv\dot{\zeta} - mgd\zeta = 0$$

to first order. Exactly as for Foucault's pendulum (§2.4) this represents epicyclic motion with frequencies

$$\Omega = \frac{1}{2}\left(\frac{Cv}{A} \pm p\right) \quad \text{where } p^2 = \frac{C^2v^2}{A^2} - \frac{4mgd}{A}.$$

This p^2 coincides with (9) evaluated at zero inclination, and the two values Ω are the possible rates of steady precession near the vertical. We conclude that a sleeping top is unstable when spinning so slowly that p is not real (since then ζ contains an unbounded time factor inconsistent with the trial assumption of perpetually small motion). On the other hand, when the top spins so rapidly that $p > 0$ the linearized solution no more than indicates the likelihood of stability. To prove it conclusively we must return to the exact equations. If the disturbance is such as to make $D \neq Cv/A$ it has been seen that the orbit is contained in a vanishingly small annulus about the steady state given by the second of (7):

$$p\alpha^2 \simeq 2\left|D - \frac{Cv}{A}\right|, \quad p > 0,$$

as $D \to Cv/A$. The corresponding epicyclic approximation is an ellipse described with frequency $\frac{1}{2}p$ (half that in θ) relative to a frame turning about the vertical at rate $Cv/2A$. If the disturbance leaves $D = Cv/A$ the orbit is a succession of loops through the pole; these are contained within a vanishingly small cap when $p \geqslant 0$ since

$$\Psi(\theta) - \Psi(0) = \frac{C^2v^2}{2A^2}\tan^2\left(\frac{\theta}{2}\right) - \frac{2mgd}{A}\sin^2\left(\frac{\theta}{2}\right)$$

is locally positive-definite. The epicyclic approximation is now a linear segment in the rotating frame, described with frequency $\sqrt{\Psi''(0)} = \frac{1}{2}p$ [whereas $\sqrt{\Psi''(\alpha)} \to p$ as $\alpha \to 0$].

(ii) *Directional gyroscopes*

When a gyrostat is mounted so that it can rotate freely about its mass-centre the earth's rotation can no longer be ignored. In the absence of any net couple the axis of symmetry would precess steadily about, or stay coincident with, an invariable direction in a sidereal frame. By means of such gyroscopic experiments a terrestrial observer could in principle determine the relative orientation of the earth's axis and hence his own latitude and the direction of true north. However, as proposed by Foucault, this information is more easily obtained with a gyrostat whose axis is smoothly constrained to move in a plane fixed to the earth.

Suppose this plane is the local horizontal in latitude λ. Let θ be the inclination of the positive C axis of symmetry to the west of north. The axial component of spin relative to a *sidereal* frame is denoted by ν. Since no axial couple, real or fictitious, is associated with this frame ν is constant, as shown previously. The variation of θ is determined by the vanishing of the vertical component of the sidereal rate of change of moment of momentum about the mass-centre. The vertical component of the moment of momentum itself is $A(\dot{\theta} + \Omega \sin \lambda)$ where Ω is the constant spin of the earth, and so the component relative rate of change is $A\ddot{\theta}$ simply. The additional convective contribution arises from the precession of the C axis about the geographical meridian, due to the horizontal component $\Omega \cos \lambda$ of the earth's spin. In all, therefore,

$$A\ddot{\theta} + (C\nu - A\Omega \cos \lambda \cos \theta)\, \Omega \cos \lambda \sin \theta = 0$$

by analogy with (2) above or with (6) in §3.8. The gyroscope can consequently remain at rest relative to the earth when aligned with the meridian ($\theta = 0$ or π). But it is stable only when the spin vector points to the north ($\theta = 0$), in which case the frequency of small oscillations is $\sqrt{(C\nu\Omega \cos \lambda/A)}$ very nearly, since $\nu \gg \Omega$, and so indicates the latitude.

Suppose, alternatively, that the plane of constraint is the meridian and let θ be the northward deviation of the positive C axis from the upward vertical. There is now no couple about the horizontal from west to east. The component moment of momentum is $-A\dot{\theta}$ while the convective contribution arises from precession about the polar axis with rate Ω and instantaneous inclination $\frac{1}{2}\pi - (\lambda + \theta)$. Whence

$$-A\ddot{\theta} + \{Cv - A\Omega \sin (\lambda + \theta)\}\Omega \cos (\lambda + \theta) = 0.$$

Stable relative equilibrium is possible only when the gyroscope is parallel to the polar axis ($\theta = \frac{1}{2}\pi - \lambda$), the oscillation frequency being $\sqrt{(Cv\Omega/A)}$, near enough. In the 'barogyroscope' a weight mg is suspended from an axial point at distance d below the mass-centre, with $mgd \gg Cv\Omega$ usually. The equilibrium deviation from the vertical is then such that

$$Cv\Omega \cos (\lambda + \theta) \simeq mgd \sin \theta \quad \text{or} \quad \theta \simeq Cv\Omega \cos \lambda/mgd.$$

If, instead, the weight is in effect suspended from a point at distance d laterally from the mass-centre, so that it exerts a restoring moment $mgd \cos \theta$ when the gyroscope is tilted above the northern horizon, the equilibrium elevation is $Cv\Omega \sin \lambda/mgd$. In both cases the frequency is $\sqrt{(mgd/A)}$.

Finally, suppose that the plane of constraint is vertical and at right angles to the meridian. Let θ be the easterly deviation of the positive C axis from the upward vertical. The component moment of momentum about the meridian is $A(\dot{\theta} + \Omega \cos \lambda)$ and the convective contribution is due to precession about the vertical with rate $\Omega \sin \lambda$ and instantaneous inclination θ. Then

$$A\ddot{\theta} + (Cv - A\Omega \sin \lambda \cos \theta)\Omega \sin \lambda \sin \theta = 0,$$

which leads to an oscillation frequency $\sqrt{(Cv\Omega \sin \lambda/A)}$ about the upward vertical.

For an idealized analysis of the modern gyrocompass see Exercise 26.

EXERCISES

1. Two intersecting axes are respectively specified by a fixed unit vector $\mathbf{k}$ and a varying (unit) vector $\mathbf{l}(t)$. Show that the plane containing them has a component spin $(\mathbf{k}\mathbf{l}\dot{\mathbf{l}})/(\mathbf{k} \wedge \mathbf{l})^2$ about the fixed axis.

2. Two planets with different inclinations to the ecliptic have constant spins $\mathbf{\Omega}_1$ and $\mathbf{\Omega}_2$ in a sidereal frame. Explain qualitatively why, to an astronomer on either planet, their relative spin does not appear constant in his local frame. State its apparent rate of change in terms of $\mathbf{\Omega}_1$ and $\mathbf{\Omega}_2$.

3. A sphere rolls between parallel planes, whose spins are ω_1 and ω_2, respectively, about separate axes along fixed normals. Show that these constraints determine the velocity of the centre G as $\mathbf{n} \wedge \frac{1}{2}(\omega_1 \mathbf{r}_1 + \omega_2 \mathbf{r}_2)$, when $\mathbf{n}$ is the unit normal and $\mathbf{r}_1$ and $\mathbf{r}_2$ are drawn towards G from any origins on the two axes of rotation. How far do the constraints determine the spin of the sphere? Prove that G describes a circle in the mid-plane and state the centre and period.

4. A body of arbitrary shape rolls on a fixed plane. Their distance apart near the point of contact P can be expressed as $z = \frac{1}{2}(\kappa_1 x_1^2 + \kappa_2 x_2^2) + \ldots$, where x_1 and x_2 are coordinates along the orthogonal lines of curvature on the body surface and κ_1 and κ_2 are the local principal curvatures, both positive. If the instantaneous body spin (ω_1, ω_2) is entirely about an axis parallel to the plane, show that the local tangent to the path of the point P is the conjugate diameter of the ellipse $z =$ constant. Show further that the acceleration of the particle of the body momentarily at P is $\omega_1^2/\kappa_2 + \omega_2^2/\kappa_1$ along the inward normal.

5. The moment of momentum of a rigid body about one of its particles P is given zero. With the notation of §3.2 show that

$$2T = m\mathbf{u}^2(P) - \boldsymbol{\omega}\mathbf{h}_{\mathrm{rel}}(P)$$

by two distinct methods.

6. Suppose that an instantaneous axis, not necessarily fixed, exists for a considered rigid motion of a body of mass m. Express the work principle (§3.2) in terms of quantities relating to this axis. Deduce that the operative forces have a component moment about the axis equal to $mkd(k\omega)/dt$, where $k(t)$ is the appropriate radius of gyration and $\omega(t)$ the component spin.

7. The equations of motion of an arbitrary continuum are formulated in a frame α whose spin is $\boldsymbol{\Omega}(t)$ in a background frame β. Obtain the mass integrals representing the overall moments about the mass-centre G of the respective fictitious body forces for α *vis-à-vis* β. Reduce the moments of the centrifugal and Euler forces to $-\boldsymbol{\Omega} \wedge I\boldsymbol{\Omega}$ and $-I\dot{\boldsymbol{\Omega}}$ respectively, where I is the inertia tensor at G [cf. the derivation of (18) in §3.2].

When the continuum is rigid, reduce the moment of the Coriolis forces to $2\boldsymbol{\omega}_{\mathrm{rel}} \wedge (J - I)\boldsymbol{\Omega}$ where $\boldsymbol{\omega}_{\mathrm{rel}}$ is the body spin relative to α and J is the polar moment of inertia about G. In this spirit interpret Euler's equations (§3.6) from the viewpoint of an *embedded* frame α.

8. A sphere rolls under gravity in a fixed spherical container (§3.3). Investigate the frictional forces in the following manner. (i) Verify the kinematical connexion $\mathbf{n} \wedge \mathbf{v}_G = a(\boldsymbol{\omega} - \nu\mathbf{n})$ and hence convert the tangential component of the linear equation of balance to

$$a(\dot{\boldsymbol{\omega}} - \nu\dot{\mathbf{n}}) = \mathbf{n} \wedge (\mathbf{g} + \mathbf{c}).$$

(ii) By combining with the rotational equation of balance, derive

$$\mathbf{n} \wedge \left\{\left(1 + \frac{a^2}{k^2}\right)\mathbf{c} + \mathbf{g}\right\} = -a\nu\dot{\mathbf{n}}.$$

(iii) Decompose this into meridional and transverse components. Show that, in a given motion, the meridional friction depends only on the angular coordinate θ, while the transverse friction depends only on $\dot{\theta}$.

9. Investigate the orbit of a sphere of mass m on a fixed horizontal plane where the available frictional force can not exceed mf in amount. Suppose the initial speed and spin given to the sphere are such it slides to begin with. Assume that the friction is $-mf\mathbf{l}$ where $u\mathbf{l}$ is the slipping velocity of the particle at the point of contact P. Show that

$$\frac{d}{dt}(u\mathbf{l}) + \left(1 + \frac{a^2}{k^2}\right) f\mathbf{l} = 0.$$

Deduce that $\mathbf{l}$ is an unvarying direction and that slipping ceases after a finite interval during which the orbit is parabolic. What is the motion thereafter?

10. A rigid body has spin $\boldsymbol{\omega}$ relative to coordinate axes x_j on which the time-dependent components of the inertia tensor I at a given particle are I_{ij} $(i, j = 1, 2, 3)$. Starting from the definition in §3.4, equation (3), obtain their rates of change in the form

$$\frac{dI_{ij}}{dt} = I_{ik}\omega_{jk} + I_{jk}\omega_{ik}.$$

Here ω_{ij} is the antisymmetric tensor defined in (2) of §3.1, with the property that the jth component of $\boldsymbol{\omega} \wedge \mathbf{a}$ can be written as $\omega_{jk}a_k$, where $\mathbf{a}$ is any vector, in particular the spin itself or the moment of momentum. Hence verify the observation in §3.2 that $(dI/dt)\boldsymbol{\omega} = \boldsymbol{\omega} \wedge I\boldsymbol{\omega}$.

11. An arbitrary line through the mass-centre G of a continuum is specified by the vector $\mathbf{l}$. Show that any parallel line in the plane of $\mathbf{l}$ and $I(G)\mathbf{l}$ is a principal direction at some point and that the product of inertia for direction $\mathbf{l}$ and the plane normal vanishes at every point [by using (13) in §3.4 or otherwise]. By taking coordinate axes at G along $\mathbf{l}$ and the plane normal, prove that the locus of such points is generally a rectangular hyperbola and discuss exceptions.

12. Establish a relation between the inertia tensors at any pair of points P and Q in a continuum, namely

$$I_{ij}(P) - I_{ij}(Q) = m\{(\xi_i x_j + \xi_j x_i) - 2(\boldsymbol{\xi}\mathbf{x})\delta_{ij}\}$$

where $\boldsymbol{\xi}$ is the vector from the mass-centre G to the midpoint of PQ and $\mathbf{x}$ is the vector from P to Q. Verify that the bisectors of the angles between these vectors are principal axes of the tensor difference. Deduce the following general theorems.

(i) If PQ is principal at P and Q it must contain G and be principal everywhere.

(ii) If the principal directions at P and Q are all parallel, but none are coincident, either G must be the midpoint of PQ or the plane GPQ must be principal everywhere.

13. A body with an axis of kinetic symmetry is initially at rest and unconstrained. An arbitrary impulse is applied at a point on this axis. Investigate all aspects of the ensuing motion, on the lines of §3.5 (i) and (ii), and locate the instantaneous axis.

14. An arbitrary body is smoothly pivoted at a fixed point, where the inertia tensor and its inverse are I and J (§3.5). When the body is at rest given impulses are applied, with total moment $\mathbf{h}$ about the pivot. Show that the kinetic energy is in general greater than when the body is smoothly mounted so that it can only rotate about some fixed axis through the pivot. In preparation prove the Cauchy-type inequality $(\mathbf{x}I\mathbf{x})(\mathbf{y}J\mathbf{y}) \geqslant (\mathbf{x}\mathbf{y})^2$ by considering the sign of the scalar product of vectors $\lambda I\mathbf{x} - \mathbf{y}$ and $\lambda\mathbf{x} - J\mathbf{y}$ for any λ.

15. An idealized rod, possibly with non-uniform density, has a radius of gyration k about any transverse axis through its midpoint, which is also the mass-centre. Its motion is specified by the velocities $\mathbf{v}_1$ and $\mathbf{v}_2$ of its ends, and its location by the directed segment $2\mathbf{a}$. Show that the kinetic energy is

$$\tfrac{1}{2}m\{\alpha(\mathbf{v}_1^2 + \mathbf{v}_2^2) + 2\beta\mathbf{v}_1\mathbf{v}_2\}$$

where $\alpha, \beta = \tfrac{1}{4}(1 \pm k^2/a^2)$. Hence, or otherwise, derive the relations

$$\mathbf{L}_1 = m(\alpha\mathbf{v}_1 + \beta\mathbf{v}_2 + \lambda\mathbf{a}), \qquad \mathbf{L}_2 = m(\beta\mathbf{v}_1 + \alpha\mathbf{v}_2 - \lambda\mathbf{a}),$$

when the motion is due to impulses $\mathbf{L}_1$ and $\mathbf{L}_2$ applied at the ends. Explain the presence of the arbitrary scalar λ when the velocities are given, and evaluate it when the impulses are given.

Several such rods are freely jointed end-to-end in a non-linear chain. Show that the motion generated by jerking one end of the chain is determinable from difference equations of type

$$\beta\mathbf{v}_{r-1} + 2\alpha\mathbf{v}_r + \beta\mathbf{v}_{r+1} = \lambda_r\mathbf{a}_r - \lambda_{r+1}\mathbf{a}_{r+1}$$

where $2\mathbf{a}_r$ is a typical segment with terminal velocities $\mathbf{v}_{r-1}$ and $\mathbf{v}_r$.

16. A body pivoted at its mass-centre G is subject to a couple $\mathbf{a} \wedge \boldsymbol{\omega}$ where $\mathbf{a}$ has constant components (a, b, c) on the principal axes at G, the moments of inertia being A, B, C. Show that in a steady state the spin is parallel to $\mathbf{a} + \mathbf{h}$ where $\mathbf{h}$ is the moment of momentum. Obtain the linearized Euler equations of disturbed motion (§3.6) and hence prove the steady spin

$$\left(\frac{a}{\lambda - A}, \frac{b}{\lambda - B}, \frac{c}{\lambda - C}\right)$$

unstable when the parameter λ makes

$$\Sigma Aa^2(\lambda - B)^3 (\lambda - C)^3 < 0.$$

17. A body is smoothly mounted on a fine shaft through the mass-centre G, where A, B, C are the principal moments of inertia. The shaft is in the B direction and can turn freely about a fixed perpendicular axis through G. No moment other than the reaction of the shaft acts on the body. Obtain and interpret the first integrals

$$(C \cos^2 \theta + A \sin^2 \theta)\dot{\phi} = \text{constant},$$
$$(C \cos^2 \theta + A \sin^2 \theta)\dot{\phi}^2 + B\dot{\theta}^2 = \text{constant},$$

where θ is the inclination of the C axis to z and ϕ is the azimuth measured round z. By Euler's differential equations, or otherwise, show that the reactive couple is

$$\left\{\frac{C \cos^2 \theta - A \sin^2 \theta}{C \cos^2 \theta + A \sin^2 \theta}(C - A) - B\right\} \dot{\theta}\dot{\phi}$$

and state its axis. Prove that the rotation about the shaft can be either monotonic or oscillatory.

18. For a body rotating freely about its mass-centre show that the quadratic form $\boldsymbol{\omega}K\boldsymbol{\omega}$ in the spin components is constant in time when $K = \lambda I - I^2$, where I is the local inertia tensor and λ is any scalar.

A material ellipsoid, geometrically similar to the energy ellipsoid and pivoted at its centre O, is in rolling contact with a flat surface positioned similarly to the fixed plane in Poinsot's representation (§3.7). Show that the two motions are identical in space and time when O is the mass-centre of the material ellipsoid; the inertia tensor there is cK for any scalar c of appropriate sign and any λ not between the least and greatest principal values of I; no forces act except at O and the point of contact; the initial orientation and energy are suitably chosen. If, additionally, the material ellipsoid is required to be homogeneous, show that λ must be the sum of the principal values of I (Sylvester 1866).

Prove that the reaction couple about O is

$$c\{I(\boldsymbol{\omega} \wedge \mathbf{h}) - \boldsymbol{\omega} \wedge I\mathbf{h}\}, \qquad \mathbf{h} = I\boldsymbol{\omega}$$

independently of λ.

19. A uniform rectangular lamina is unconstrained and is initially spinning about a diagonal at rate n. Verify that $h^2 = CT$ and that the other diagonal is initially parallel to the invariable plane. Prove that the body cone encloses the longer geometric axis of symmetry and is described in period

$$\frac{4}{n}\int_0^{\alpha}\left(\frac{\sec 2\alpha}{\sin^2\alpha - \sin^2\chi}\right)^{\frac{1}{2}} d\chi$$

where 2α is the acute angle between the diagonals and where χ is as in §3.8(ii). What happens when the lamina is square?

20. In the free rotation of an axially symmetric body about its mass-centre show, with the notation of §3.8:

(i) $h^2/2T = AC/I$ where I is the moment of inertia about a specified axis perpendicular to the invariable line.

(ii) When $C > A$ the semi-angle of the space cone can not exceed $\sin^{-1}(\frac{1}{3}) = 19°28'$, this being attained when the body is a disc set spinning about a direction at $\tan^{-1}\sqrt{2} = 54°44'$ to the normal. Verify that $\alpha + \beta = \frac{1}{2}\pi$ and draw the space and body cones.

21. A heavy top is pivoted at a fixed point on its axis. Show that an additional couple with magnitude

$$A(\Omega - \Omega_1)(\Omega - \Omega_2)\sin\alpha\cos\alpha$$

must be applied to maintain an arbitrary rate of precession Ω, where Ω_1 and Ω_2 are the rates of free precession at the same inclination α.

22. A gyrostat has a spindle PQ of length $2a$, the mass-centre is at the mid-point G, the moments of inertia there are A and C, and its mass is m. The gyrostat precesses steadily round the vertical at rate Ω, the axial component of spin is ν, and α is the angle included by their right-handed directions. Obtain the following relations under various constraints, using the virtual work-rate principle where convenient.

(i) If P and Q move freely along the arms of a light right-angled guide, one of which is fixed vertically and the other free to rotate,

$$(A + ma^2)\Omega^2\cos\alpha = C\nu\Omega + mga.$$

(ii) If P is made to describe a horizontal circle while Q is free,

$$(A + mar\operatorname{cosec}\alpha)\Omega^2\cos\alpha = C\nu\Omega + mga,$$

where r is the orbital radius of G.

(iii) If P is attached to a fixed point by a light string of length l and inclination β, Q being free,

$$A\Omega^2 \cos \alpha - Cv\Omega = mga\,(1 - \tan \beta \cot \alpha),$$
$$(a \sin \alpha + l \sin \beta)\Omega^2 = g \tan \beta.$$

23. At its lowest level the axis of a heavy top is horizontal and turning round the vertical at rate $\Omega = 2mgd/Cv$ (notation of §3.9). Discuss the character and range of the orbit according to the sign of $A\Omega/Cv - 1$.

24. Prove that a sleeping top with spin v less than $\omega = 2\sqrt{(Amgd)}/C$ is unstable by showing that the axis will attain an inclination $2 \cos^{-1} (v/\omega)$, very nearly, under certain vanishingly small disturbances (notation of §3.9).

A homogeneous cube of side $2a$ is set spinning about a vertical long diagonal whose lowest point is a fixed pivot. Show that the motion is unstable when the spin is less than $\{33\sqrt{(3)}g/a\}^{\frac{1}{2}}$.

25. The rate of precession of a top is maintained at a constant value Ω by applying a suitable horizontal force perpendicular to the axis. Show that a steady inclination γ is possible when $|Cv\Omega - mgd| < A\Omega^2$ and that the frequency of a small oscillation is then $\Omega \sin \gamma$. If the inclination ranges between α and β show that

$$\cos \alpha + \cos \beta = 2(Cv\Omega - mgd)/A\Omega^2.$$

26. A modern gyrocompass essentially consists of a gyrostat which can rotate freely about its mass-centre. A unit vector along its C axis of symmetry is denoted by $\mathbf{l}$, with tilt ε above the horizon and deviation δ east of north. A weight is attached to the gimbals in such a way as to produce a moment $mgd \sin \varepsilon$ opposing tilting. Obtain the equation of balance in a terrestrial frame, and at a fixed station, in the form

$$A\mathbf{l} \wedge \ddot{\mathbf{l}} + Cv(\dot{\mathbf{l}} + \boldsymbol{\Omega} \wedge \mathbf{l}) = \mathbf{l} \wedge mgd \sin \varepsilon,$$

neglecting terms in the earth's spin $\boldsymbol{\Omega}$ except when it is in product with the component axial spin v of the gyrostat.

If $mgd \gg Cv\Omega$ show that the axis can remain at relative rest when pointing north and tilted slightly, and that in small oscillations about this position

$$A\ddot{\delta} - Cv\dot{\varepsilon} + Cv\delta\Omega \cos \lambda = 0,$$
$$A\ddot{\varepsilon} + Cv\dot{\delta} + mgd\varepsilon = Cv\Omega \sin \lambda,$$

where λ is the latitude. Deduce that the principal frequencies are Cv/A and $\sqrt{(mgd\,\Omega \cos \lambda/Cv)}$ approximately, when $C^2v^2/A \gg mgd$.

27. A continuum, not necessarily rigid, has an inertia tensor I at its mass-centre G. The motion of an adopted frame α relative to a background frame β is an instantaneous spin $\boldsymbol{\Omega}$ about G. The kinetic energies of the continuum relative to the two frames are T_α and T_β, while the moments of momentum about G are $\mathbf{h}_\alpha$ and $\mathbf{h}_\beta$. Obtain the connexions

$$\mathbf{h}_\beta - \mathbf{h}_\alpha = I\boldsymbol{\Omega}, \qquad T_\beta - T_\alpha = \tfrac{1}{2}\boldsymbol{\Omega}(\mathbf{h}_\alpha + \mathbf{h}_\beta).$$

Show that T_α is an absolute minimum for arbitrary choice of $\boldsymbol{\Omega}$ when $\mathbf{h}_\alpha = \mathbf{0}$. [Coordinates in such an α are known as 'Tisserand's mean axes', and are used for instance in theories of bodily tides and other deformations of the earth.]

28. Show, by means of the divergence theorem, that the volume averages of the components of stress in a continuum of volume V and arbitrary material properties are such that

$$V\bar{\sigma}_{ij} = \int x_i \tau_j dS + \int x_i(\gamma_j - \dot{v}_j)dm = \int x_j \tau_i dS + \int x_j(\gamma_i - \dot{v}_i)dm$$

in the notation of §2.11 and with the help of (2) and (3) there.

A rigid body is subject neither to surface traction nor body force. Evaluate the field of acceleration in terms of the spin $\boldsymbol{\omega}$, having regard to Euler's equations (§3.6). Hence prove that

$$V\bar{\sigma}_{11} = (\omega_2^2 + \omega_3^2)\int x_1^2 dm, \qquad V\bar{\sigma}_{12} = -2\omega_1\omega_2 \int x_1^2 dm \int x_2^2 dm/C,$$

where $\int x_1^2 dm = \frac{1}{2}(B + C - A)$, etc.,

the coordinates being the principal axes of inertia at the mass-centre. Reduce these expressions when the body is axially symmetric and precesses.

Investigate also the case when a rigid body is freely pivoted at any fixed point and subject only to its weight and the reaction.

INDEX